Geographic Information Systems

Geographic Information Systems

Colby Blair

NY RESEARCH PRESS
New York

Published by NY Research Press
118-35 Queens Blvd., Suite 400,
Forest Hills, NY 11375, USA
www.nyresearchpress.com

Geographic Information Systems
Colby Blair

International Standard Book Number: 978-1-63238-700-4 (Hardback)

Cataloging-in-Publication Data

Geographic information systems / Colby Blair.
p. cm.
Includes bibliographical references and index.
ISBN 978-1-63238-700-4
1. Geographic information systems. 2. Information storage and retrieval systems--Geography.
I. Blair, Colby.
G70.212 .G46 2019
910.285--dc23

Printed in China.

Contents

Preface

A geographic information system (GIS) is a modern technology that is designed to capture, analyze, store and manipulate geographic or spatial data. GIS can be used for editing data in maps, creating interactive queries, analyzing spatial information, etc. Such applications of GIS enable location-enabled services, which depend on visualization and analysis. Such services are of use in diverse areas of science, engineering, defense, sustainable development, transport, climatology, telecommunications, etc. GIS can be applied for data analysis, hydrological modeling, topological modeling, data output and cartography, among others. This book elucidates the concepts and innovative models around prospective developments with respect to geographic information systems. It elucidates new techniques and their applications in a multidisciplinary manner. In this book, constant effort has been made to make the understanding of the difficult concepts of geographic information systems, as easy and informative as possible, for the readers.

To facilitate a deeper understanding of the contents of this book a short introduction of every chapter is written below:

Chapter 1- A geographic information system or GIS is used for capturing, analyzing, storing, manipulating and presenting geographic and spatial data. This is an introductory chapter which will introduce briefly all the significant aspects of GIS, as well as its different types such as four dimensional GIS, web GIS, virtual reality GIS and multimedia GIS.

Chapter 2- Remote sensing enables the acquisition of information about an event, object or phenomenon without making any direct contact with the area or object under observation. This chapter includes valuable insights into significant topics like water remote sensing, multispectral imaging, normalized difference vegetation index and GIS in environmental contamination, which are crucial for a comprehensive understanding of techniques and applications of GIS.

Chapter 3- In geometry, the coordinate system that helps in determining and locating every location on Earth, by specifying a unique set of letters, numbers or symbols, is termed as a geographic coordinate system. The coordinates are so chosen so as to represent the location's vertical and horizontal position. This chapter closely examines the different geographic coordinate systems such as geodetic datum and the Cartesian coordinate system.

Chapter 4- The science that is concerned with the practice of making maps is called cartography. It is based on the premise that reality can be represented in a way that effectively communicates spatial information. This chapter analyzes the fundamental concepts of terrain cartography, cartographic principles and design, map projection, etc. to provide an easy understanding of cartography.

Chapter 5- A geographical database is a database that stores geographical data pertaining to administrative divisions, cities and countries. This chapter has been carefully written to provide an easy understanding of the varied facets of geographical databases, by incorporating several significant topics such as aerodrome mapping database, barrow area information database, European soil database, etc.

Chapter 6- Encoding geographical information into a computer file requires a specific standard, which is called a GIS file format. The understanding of Raster graphics and Vector graphics is vital to the area of GIS file formats. This chapter discusses in detail some of the popular GIS file formats, such as GIS Raster formats and GIS Vector formats.

Finally, I would like to thank the entire team involved in the inception of this book for their valuable time and contribution. This book would not have been possible without their efforts. I would also like to thank my friends and family for their constant support.

Colby Blair

Introduction to Geographic Information System

A geographic information system or GIS is used for capturing, analyzing, storing, manipulating and presenting geographic and spatial data. This is an introductory chapter which will introduce briefly all the significant aspects of GIS, as well as its different types such as four dimensional GIS, web GIS, virtual reality GIS and multimedia GIS.

A Geographical Information System (GIS) is a system for capturing, storing, analyzing and managing data and associated attributes, which are spatially referenced to the Earth. The geographical information system is also called as a geographic information system or geospatial information system. It is an information system capable of integrating, storing, editing, analyzing, sharing, and displaying geographically referenced information. In a more generic sense, GIS is a software tool that allows users to create interactive queries, analyze the spatial information, edit data, maps, and present the results of all these operations. GIS technology is becoming essential tool to combine various maps and remote sensing information to generate various models, which are used in real time environment. Geographical information system is the science utilizing the geographic concepts, applications and systems.

Geographical Information System can be used for scientific investigations, resource management, asset management, environmental impact assessment, urban planning, cartography, criminology, history, sales, marketing, and logistics. For example, agricultural planners might use geographical data to decide on the best locations for a location specific crop planning, by combining data on soils, topography, and rainfall to determine the size and location of biologically suitable areas. The final output could include overlays with land ownership, transport, infrastructure, labour availability, and distance to market centers.

Geographical Information System (GIS) is a technology that provides the means to collect and use geographic data to assist in the development of Agriculture. A digital map is generally of much greater value than the same map printed on a paper as the digital version can be combined with other sources of data for analyzing information with a graphical presentation. The GIS software makes it possible to synthesize large amounts of different data, combining different layers of information to manage and retrieve the data in a more useful manner. GIS provides a powerful means for agricultural scientists to better service to the farmers and farming community in answering their query and helping in a better decision making to implement planning activities for the development of agriculture.

Components of GIS

GIS enables the user to input, manage, manipulate, analyze, and display geographically referenced data using a computerized system. To perform various operations with GIS, the components of GIS such as software, hardware, data, people and methods are essential.

Software

GIS software provides the functions and tools needed to store, analyze, and display geographic information. Key software components are (a) a database management system (DBMS) (b) tools for the input and manipulation of geographic information (c) tools that support geographic query, analysis, and visualization (d) a graphical user interface (GUI) for easy access to tools. GIS software are either commercial software or software developed on Open Source domain, which are available for free. However, the commercial software is copyright protected, can be expensive and is available in terms number of licensees.

Currently available commercial GIS software includes Arc/Info, Intergraph, MapInfo, Gram++ etc. Out of these Arc/Info is the most popular software package. And, the open source software are AMS/MARS etc.

Hardware

Hardware is the computer on which a GIS operates. Today, GIS runs on a wide range of hardware types, from centralized computer servers to desktop computers used in stand-alone or networked configurations. Minimum configuration required to Arc/Info Desktop 9.0 GIS application is as follows:

Product: ArcInfo Desktop 9.0

Platform: PC-Intel

Operating System: Windows XP Professional Edition, Home Edition

Service Packs/Patches: SP 1

SP2 (refer to Limitations)

Shipping/Release Date: May 10, 2004

Hardware Requirements

CPU Speed:	800 MHz minimum, 1.0 GHz recommended or higher
Processor:	Pentium or higher
Memory/RAM:	256 MB minumum, 512 MB recommended or higher
Display Properties:	Greater than 256 color depth
Swap Space:	300 MB minimum
Disk Space:	Typical 605 MB NTFS, Complete 695 MB FAT32 + 50 MB for installation
Browser :	Internet Explorer 6.0 Requirement

(Some features of ArcInfo Desktop 9.0 require a minimum installation of Microsoft Internet Explorer Version 6.0.)

Data

The most important component of a GIS is the data. Geographic data or Spatial data and related tabular data can be collected in-house or bought from a commercial data provider. Spatial data can be in the form of a map/remotely-sensed data such as satellite imagery and aerial photography. These data forms must be properly georeferenced (latitude/longitude). Tabular data can be in the form attribute data that is in some way related to spatial data. Most GIS software comes with inbuilt Database Management Systems (DBMS) to create and maintain a database to help organize and manage data.

Users

GIS technology is of limited value without the users who manage the system and to develop plans for applying it. GIS users range from technical specialists who design and maintain the system to those who use it to help them do their everyday work.

These users are largely interested in the results of the analysis and may have no interest or knowledge of the methods of analysis. The user-friendly interface of the GIS software allows the non-technical users to have easy access to GIS analytical capabilities without needing to know detailed software commands. A simple User Interface (UI) can consist of menus and pull-down graphic windows so that the user can perform required analysis with a few key presses without needing to learn specific commands in detail.

Methods

A successful GIS operates according to a well-designed plan and business rules, which are the models and operating practices unique to each organization.

GIS Techniques and Technology

Modern GIS technologies use digital information, for which various digitized data creation methods are used. The most common method of data creation is digitization, where a hard copy map or survey plan is transferred into a digital medium through the use of a CAD program, and geo-referencing capabilities. With the wide availability of ortho-rectified imagery (from satellites, aircraft, Helikites and UAVs), heads-up digitizing is becoming the main avenue through which geographic data is extracted. Heads-up digitizing involves the tracing of geographic data directly on top of the aerial imagery instead of by the traditional method of tracing the geographic form on a separate digitizing tablet (heads-down digitizing).

Relating Information from Different Sources

GIS uses spatio-temporal (space-time) location as the key index variable for all other information. Just as a relational database containing text or numbers can relate many different tables using common key index variables, GIS can relate otherwise unrelated information by using location as the key index variable. The key is the location and/or extent in space-time.

Any variable that can be located spatially, and increasingly also temporally, can be referenced using a GIS. Locations or extents in Earth space–time may be recorded as dates/times of occurrence, and x, y, and z coordinates representing, longitude, latitude, and elevation, respectively. These GIS coordinates may represent other quantified systems of temporo-spatial reference (for example, film frame number, stream gage station, highway mile-marker, surveyor benchmark, building address, street intersection, entrance gate, water depth sounding, POSor CAD drawing origin/units). Units applied to recorded temporal-spatial data can vary widely (even when using exactly the same data, see map projections), but all Earth-based spatial–temporal location and extent references should, ideally, be relatable to one another and ultimately to a "real" physical location or extent in space–time.

Related by accurate spatial information, an incredible variety of real-world and projected past or future data can be analyzed, interpreted and represented.This key characteristic of GIS has begun to open new avenues of scientific inquiry into behaviors and patterns of real-world information that previously had not been systematically correlated.

GIS Uncertainties

GIS accuracy depends upon source data, and how it is encoded to be data referenced. Land surveyors have been able to provide a high level of positional accuracy utilizing the GPS-derived positions. High-resolution digital terrain and aerial imagery, powerful computers and Web technology are changing the quality, utility, and expectations of GIS to serve society on a grand scale, but nevertheless there are other source data that affect overall GIS accuracy like paper maps, though these may be of limited use in achieving the desired accuracy.

In developing a digital topographic database for a GIS, topographical maps are the main source, and aerial photography and satellite imagery are extra sources for collecting data and identifying attributes which can be mapped in layers over a location facsimile of scale. The scale of a map and geographical rendering area representation type are very important aspects since the information content depends mainly on the scale set and resulting locatability of the map's representations. In order to digitize a map, the map has to be checked within theoretical dimensions, then scanned into a raster format, and resulting raster data has to be given a theoretical dimension by a rubber sheeting/warping technology process.

A quantitative analysis of maps brings accuracy issues into focus. The electronic and other equipment used to make measurements for GIS is far more precise than the machines of conventional map analysis. All geographical data are inherently inaccurate, and these inaccuracies will propagate through GIS operations in ways that are difficult to predict.

Data Representation

GIS data represents real objects (such as roads, land use, elevation, trees, waterways, etc.) with digital data determining the mix. Real objects can be divided into two abstractions: discrete objects (e.g., a house) and continuous fields (such as rainfall amount, or elevations). Traditionally, there are two broad methods used to store data in a GIS for both kinds of abstractions mapping references: raster images and vector. Points, lines, and polygons are the stuff of mapped location attribute references. A new hybrid method of storing data is that of identifying point clouds, which combine three-dimensional points with RGB information at each point, returning a "3D color image". GIS thematic maps then are becoming more and more realistically visually descriptive of what they set out to show or determine.

Data Capture

Example of hardware for mapping (GPS and laser rangefinder) and data collection (rugged computer). The current trend for geographical information system (GIS) is that accurate mapping and data analysis are completed while in the field. Depicted hardware (field-maptechnology) is used mainly for forest inventories, monitoring and mapping.

Data capture—entering information into the system—consumes much of the time of GIS practitioners. There are a variety of methods used to enter data into a GIS where it is stored in a digital format.

Existing data printed on paper or PET film maps can be digitized or scanned to produce digital data. A digitizer produces vector data as an operator traces points, lines, and polygon boundaries from a map. Scanning a map results in raster data that could be further processed to produce vector data.

Survey data can be directly entered into a GIS from digital data collection systems on survey instruments using a technique called coordinate geometry (COGO). Positions from a global navigation satellite system (GNSS) like Global Positioning System can also be collected and then imported into a GIS. A current trend in data collection gives users the ability to utilize field computers with the ability to edit live data using wireless connections or disconnected editing sessions. This has been enhanced by the availability of low-cost mapping-grade GPS units with decimeter accuracy in real time. This eliminates the need to post process, import, and update the data in the office after fieldwork has been collected. This includes the ability to incorporate positions collected using a laser rangefinder. New technologies also allow users to create maps as well as analysis directly in the field, making projects more efficient and mapping more accurate.

Remotely sensed data also plays an important role in data collection and consist of sensors attached to a platform. Sensors include cameras, digital scanners and lidar, while platforms usually consist of aircraft and satellites. In England in the mid 1990s, hybrid kite/balloons called helikites first pioneered the use of compact airborne digital cameras as airborne geo-information systems. Aircraft measurement software, accurate to 0.4 mm was used to link the photographs and measure the ground. Helikites are inexpensive and gather more accurate data than aircraft. Helikites can be used over roads, railways and towns where unmanned aerial vehicles (UAVs) are banned.

Recently aerial data collection is becoming possible with miniature UAVs. For example, the Aeryon Scout was used to map a 50-acre area with a ground sample distance of 1 inch (2.54 cm) in only 12 minutes.

The majority of digital data currently comes from photo interpretation of aerial photographs. Soft-copy workstations are used to digitize features directly from stereo pairs of digital photographs. These systems allow data to be captured in two and three dimensions, with elevations measured directly from a stereo pair using principles of photogrammetry. Analog aerial photos must be scanned before being entered into a soft-copy system, for high-quality digital cameras this step is skipped.

Satellite remote sensing provides another important source of spatial data. Here satellites use different sensor packages to passively measure the reflectance from parts of the electromagnetic spectrum or radio waves that were sent out from an active sensor such as radar. Remote sensing collects raster data that can be further processed using different bands to identify objects and classes of interest, such as land cover.

When data is captured, the user should consider if the data should be captured with either a relative accuracy or absolute accuracy, since this could not only influence how information will be interpreted but also the cost of data capture.

After entering data into a GIS, the data usually requires editing, to remove errors, or further processing. For vector data it must be made "topologically correct" before it can be used for some advanced analysis. For example, in a road network, lines must connect with nodes at an intersection. Errors such as undershoots and overshoots must also be removed. For scanned maps, blemishes on the source map may need to be removed from the resulting raster. For example, a fleck of dirt might connect two lines that should not be connected.

Raster-to-vector Translation

Data restructuring can be performed by a GIS to convert data into different formats. For example, a GIS may be used to convert a satellite image map to a vector structure by generating lines around all cells with the same classification, while determining the cell spatial relationships, such as adjacency or inclusion.

More advanced data processing can occur with image processing, a technique developed in the late 1960s by NASA and the private sector to provide contrast enhancement, false color rendering and a variety of other techniques including use of two dimensional Fourier transforms. Since digital data is collected and stored in various ways, the two data sources may not be entirely compatible. So a GIS must be able to convert geographic data from one structure to another. In so doing, the implicit assumptions behind different ontologies and classifications require analysis. Object ontologies have gained increasing prominence as a consequence of object-oriented programming and sustained work by Barry Smithand co-workers.

Projections, Coordinate Systems, and Registration

The earth can be represented by various models, each of which may provide a different set of coordinates (e.g., latitude, longitude, elevation) for any given point on the Earth's surface. The simplest model is to assume the earth is a perfect sphere. As more measurements of the earth have accumulated, the models of the earth have become more sophisticated and more accurate. In fact, there are models called datums that apply to different areas of the earth to provide increased accuracy, like NAD83 for U.S. measurements, and the World Geodetic System for worldwide measurements.

Spatial Analysis with Geographical Information System (GIS)

GIS spatial analysis is a rapidly changing field, and GIS packages are increasingly including analytical tools as standard built-in facilities, as optional toolsets, as add-ins or 'analysts'. In many instances these are provided by the original software suppliers (commercial vendors or collaborative non commercial development teams), while in other cases facilities have been developed and are provided by third parties. Furthermore, many products offer software development kits (SDKs), programming languages and language support, scripting facilities and/or special interfaces for developing one's own analytical tools or variants. The increased availability has created a new dimension to business intelligence termed "spatial intelligence" which, when openly delivered via intranet, democratizes access to geographic and social network data. Geospatial intelligence, based on GIS spatial analysis, has also become a key element for security. GIS as a whole can be described as conversion to a vectorial representation or to any other digitisation process.

Slope and Aspect

Slope can be defined as the steepness or gradient of a unit of terrain, usually measured as an angle in degrees or as a percentage. Aspect can be defined as the direction in which a unit of terrain faces. Aspect is usually expressed in degrees from north. Slope, aspect, and surface curvature in terrain analysis are all derived from neighborhood operations using elevation values of a cell's adjacent neighbours. Slope is a function of resolution, and the spatial resolution used to calculate slope and aspect should always be specified.Various authors have compared techniques for calculating slope and aspect.

The following method can be used to derive slope and aspect:

The elevation at a point or unit of terrain will have perpendicular tangents (slope) passing through the point, in an east-west and north-south direction. These two tangents give two components, $\partial z/\partial x$ and $\partial z/\partial y$, which then be used to determine the overall direction of slope, and the aspect of the slope. The gradient is defined as a vector quantity with components equal to the partial derivatives of the surface in the x and y directions.

The calculation of the overall 3x3 grid slope S and aspect A for methods that determine east-west and north-south component use the following formulas respectively:

$$\tan S = \sqrt{\left(\frac{\partial z}{\partial x}\right)^2 + \left(\frac{\partial z}{\partial y}\right)^2}$$

$$\tan A = \left(\frac{\left(\frac{-\partial z}{\partial y}\right)}{\left(\frac{\partial z}{\partial x}\right)}\right)$$

Zhou and Liu describe another formula for calculating aspect, as follows:

$$A = 270^{\circ} + \arctan\left(\frac{\left(\frac{\partial z}{\partial x}\right)}{\left(\frac{\partial z}{\partial y}\right)}\right) - 90^{\circ}\left(\frac{\left(\frac{\partial z}{\partial y}\right)}{\left|\frac{\partial z}{\partial y}\right|}\right).$$

Data Analysis

It is difficult to relate wetlands maps to rainfall amounts recorded at different points such as airports, television stations, and schools. A GIS, however, can be used to depict two- and three-dimensional characteristics of the Earth's surface, subsurface, and atmosphere from information points. For example, a GIS can quickly generate a map with isopleth or contour lines that indicate differing amounts of rainfall. Such a map can be thought of as a rainfall contour map. Many sophisticated methods can estimate the characteristics of surfaces from a limited number of point measurements. A two-dimensional contour map created from the surface modeling of rainfall point measurements may be overlaid and analyzed with any other map in a GIS covering the same area. This GIS derived map can then provide additional information - such as the viability of water powerpotential as a renewable energy source. Similarly, GIS can be used to compare other renewable energy resources to find the best geographic potential for a region.

Additionally, from a series of three-dimensional points, or digital elevation model, isopleth lines representing elevation contours can be generated, along with slope analysis, shaded relief, and other elevation products. Watersheds can be easily defined for any given reach, by computing all of the areas contiguous and uphill from any given point of interest. Similarly, an expected thalweg of where surface water would want to travel in intermittent and permanent streams can be computed from elevation data in the GIS.

Topological Modeling

A GIS can recognize and analyze the spatial relationships that exist within digitally stored spatial data. These topological relationships allow complex spatial modelling and analysis to be performed. Topological relationships between geometric entities traditionally include adjacency (what adjoins what), containment (what encloses what), and proximity (how close something is to something else).

Geometric Networks

Geometric networks are linear networks of objects that can be used to represent interconnected features, and to perform special spatial analysis on them. A geometric network is composed of edges, which are connected at junction points, similar to graphs in mathematics and computer science. Just like graphs, networks can have weight and flow assigned to its edges, which can be used to represent various interconnected features more accurately. Geometric networks are often used to model road networks and public utility networks, such as electric, gas, and water networks. Network modeling is also commonly employed in transportation planning, hydrology modeling, and infrastructure modeling.

Hydrological Modeling

GIS hydrological models can provide a spatial element that other hydrological models lack, with the analysis of variables such as slope, aspect and watershed or catchment area. Terrain analysis is fundamental to hydrology, since water always flows down a slope. As basic terrain analysis of a digital elevation model (DEM) involves calculation of slope and aspect, DEMs are very useful for hydrological analysis. Slope and aspect can then be used to determine direction of surface runoff, and hence flow accumulation for the formation of streams, rivers and lakes. Areas of divergent flow can also give a clear indication of the boundaries of a catchment. Once a flow direction and accumulation matrix has been created, queries can be performed that show contributing or dispersal areas at a certain point. More detail can be added to the model, such as terrain roughness, vegetation types and soil types, which can influence infiltration and evapotranspiration rates, and hence influencing surface flow. One of the main uses of hydrological modeling is in environmental contamination research. Other applications of hydrological modeling include groundwater and surface water mapping, as well as flood risk maps.

Cartographic Modeling

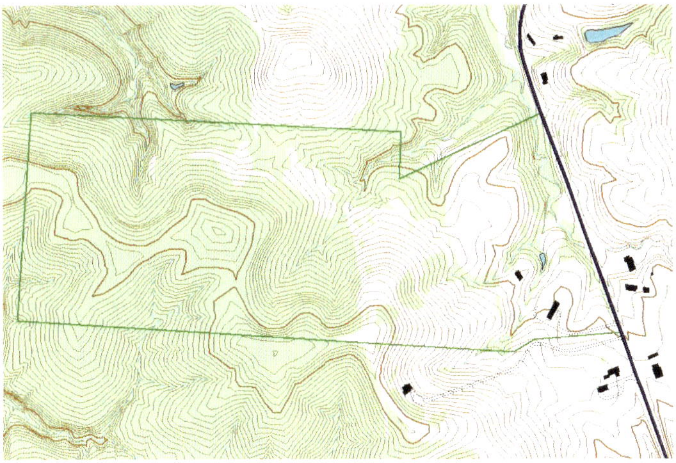

An example of use of layers in a GIS application. In this example, the forest-cover layer (light green) forms the bottom layer, with the topographic layer (contour lines) over it. Next up is a standing water layer (pond, lake) and then a flowing water layer (stream, river), followed by the boundary layer and finally the road layer on top. The order is very important in order to properly display the final result. Note that the ponds are layered under the streams, so that a stream line can be seen overlying one of the ponds.

Dana Tomlin probably coined the term "cartographic modeling" in his PhD dissertation (1983); he later used it in the title of his book, *Geographic Information Systems and Cartographic Modeling*(1990). Cartographic modeling refers to a process where several thematic layers of the same area are produced, processed, and analyzed. Tomlin used raster layers, but the overlay method can be used more generally. Operations on map layers can be combined into algorithms, and eventually into simulation or optimization models.

Map Overlay

The combination of several spatial datasets (points, lines, or polygons) creates a new output vector dataset, visually similar to stacking several maps of the same region. These overlays are similar to mathematical Venn diagram overlays. A union overlay combines the geographic features and attribute tables of both inputs into a single new output. An intersect overlay defines the area where both inputs overlap and retains a set of attribute fields for each. A symmetric difference overlay defines an output area that includes the total area of both inputs except for the overlapping area.

Data extraction is a GIS process similar to vector overlay, though it can be used in either vector or raster data analysis. Rather than combining the properties and features of both datasets, data extraction involves using a "clip" or "mask" to extract the features of one data set that fall within the spatial extent of another dataset.

In raster data analysis, the overlay of datasets is accomplished through a process known as "local operation on multiple rasters" or "map algebra", through a function that combines the values of each raster's matrix. This function may weigh some inputs more than others through use of an "index model" that reflects the influence of various factors upon a geographic phenomenon.

Geostatistics

Geostatistics is a branch of statistics that deals with field data, spatial data with a continuous index. It provides methods to model spatial correlation, and predict values at arbitrary locations (interpolation).

When phenomena are measured, the observation methods dictate the accuracy of any subsequent analysis. Due to the nature of the data (e.g. traffic patterns in an urban environment; weather patterns over the Pacific Ocean), a constant or dynamic degree of precision is always lost in the measurement. This loss of precision is determined from the scale and distribution of the data collection.

To determine the statistical relevance of the analysis, an average is determined so that points (gradients) outside of any immediate measurement can be included to determine their predicted behavior. This is due to the limitations of the applied statistic and data collection methods, and interpolation is required to predict the behavior of particles, points, and locations that are not directly measurable.

Hillshade model derived from a Digital Elevation Model of the Valestra area in the northern Apennines (Italy)

Interpolation is the process by which a surface is created, usually a raster dataset, through the input of data collected at a number of sample points. There are several forms of interpolation, each which treats the data differently, depending on the properties of the data set. In comparing interpolation methods, the first consideration should be whether or not the source data will change (exact or approximate). Next is whether the method is subjective, a human interpretation, or objective. Then there is the nature of transitions between points: are they abrupt or gradual. Finally, there is whether a method is global (it uses the entire data set to form the model), or local where an algorithm is repeated for a small section of terrain.

Interpolation is a justified measurement because of a spatial autocorrelation principle that recognizes that data collected at any position will have a great similarity to, or influence of those locations within its immediate vicinity.

Digital elevation models, triangulated irregular networks, edge-finding algorithms, Thiessen polygons, Fourier analysis, (weighted) moving averages, inverse distance weighting, kriging, spline, and trend surface analysis are all mathematical methods to produce interpolative data.

Address Geocoding

Geocoding is interpolating spatial locations (X,Y coordinates) from street addresses or any other spatially referenced data such as ZIP Codes, parcel lots and address locations. A reference theme is required to geocode individual addresses, such as a road centerline file with address ranges. The individual address locations have historically been interpolated, or estimated, by examining address ranges along a road segment. These are usually provided in the form of a table or database. The software will then place a dot approximately where that address belongs along the segment of centerline. For example, an address point of 500 will be at the midpoint of a line segment that starts with address 1 and ends with address 1,000. Geocoding can also be applied against actual parcel data, typically from municipal tax maps. In this case, the result of the geocoding will be an actually positioned space as opposed to an interpolated point. This approach is being increasingly used to provide more precise location information.

Reverse Geocoding

Reverse geocoding is the process of returning an estimated street address number as it relates to a given coordinate. For example, a user can click on a road centerline theme (thus providing a coordinate) and have information returned that reflects the estimated house number. This house number is interpolated from a range assigned to that road segment. If the user clicks at the midpoint of a segment that starts with address 1 and ends with 100, the returned value will be somewhere near 50. Note that reverse geocoding does not return actual addresses, only estimates of what should be there based on the predetermined range.

Multi-criteria Decision Analysis

Coupled with GIS, multi-criteria decision analysis methods support decision-makers in analysing a set of alternative spatial solutions, such as the most likely ecological habitat for restoration, against multiple criteria, such as vegetation cover or roads. MCDA uses decision rules to aggregate the criteria, which allows the alternative solutions to be ranked or prioritised. GIS MCDA may reduce costs and time involved in identifying potential restoration sites.

Data Output and Cartography

Cartography is the design and production of maps, or visual representations of spatial data. The vast majority of modern cartography is done with the help of computers, usually using GIS but production of quality cartography is also achieved by importing layers into a design program to refine it. Most GIS software gives the user substantial control over the appearance of the data.

Cartographic work serves two major functions:

First, it produces graphics on the screen or on paper that convey the results of analysis to the people who make decisions about resources. Wall maps and other graphics can be generated, allowing the viewer to visualize and thereby understand the results of analysis or simulations of potential events. Web Map Servers facilitate distribution of generated maps through web browsers using various implementations of web-based application programming interfaces (AJAX, Java, Flash, etc.).

Second, other database information can be generated for further analysis or use. An example would be a list of all addresses within one mile (1.6 km) of a toxic spill.

Graphic Display Techniques

Traditional maps are abstractions of the real world, a sampling of important elements portrayed on a sheet of paper with symbols to represent physical objects. People who use maps must interpret these symbols. Topographic maps show the shape of land surface with contour lines or with shaded relief.

Today, graphic display techniques such as shading based on altitude in a GIS can make relationships among map elements visible, heightening one's ability to extract and analyze information. For example, two types of data were combined in a GIS to produce a perspective view of a portion of San Mateo County, California.

- The digital elevation model, consisting of surface elevations recorded on a 30-meter horizontal grid, shows high elevations as white and low elevation as black.
- The accompanying Landsat Thematic Mapper image shows a false-color infrared image looking down at the same area in 30-meter pixels, or picture elements, for the same coordinate points, pixel by pixel, as the elevation information.

A GIS was used to register and combine the two images to render the three-dimensional perspective view looking down the San Andreas Fault, using the Thematic Mapper image pixels, but shaded using the elevation of the landforms. The GIS display depends on the viewing point of the observer and time of day of the display, to properly render the shadows created by the sun's rays at that latitude, longitude, and time of day.

An archeochrome is a new way of displaying spatial data. It is a thematic on a 3D map that is applied to a specific building or a part of a building. It is suited to the visual display of heat-loss data.

Spatial ETL

Spatial ETL tools provide the data processing functionality of traditional extract, transform, load (ETL) software, but with a primary focus on the ability to manage spatial data. They provide GIS users with the ability to translate data between different standards and proprietary formats, whilst geometrically transforming the data en route. These tools can come in the form of add-ins to existing wider-purpose software such as spreadsheets.

GIS Data Mining

GIS or spatial data mining is the application of data mining methods to spatial data. Data mining, which is the partially automated search for hidden patterns in large databases, offers great potential benefits for applied GIS-based decision making. Typical applications include environmental monitoring. A characteristic of such applications is that spatial correlation between data measurements require the use of specialized algorithms for more efficient data analysis.

Applications

The implementation of a GIS is often driven by jurisdictional (such as a city), purpose, or application requirements. Generally, a GIS implementation may be custom-designed for an organization. Hence, a GIS deployment developed for an application, jurisdiction, enterprise, or purpose may not be necessarily interoperable or compatible with a GIS that has been developed for some other application, jurisdiction, enterprise, or purpose.

GIS provides, for every kind of location-based organization, a platform to update geographical data without wasting time to visit the field and update a database manually. GIS when integrated with other powerful enterprise solutions like SAP and the Wolfram Language helps creating powerful decision support system at enterprise level.

Many disciplines can benefit from GIS technology. An active GIS market has resulted in lower costs and continual improvements in the hardware and software components of GIS, and usage in the fields of science, government, business, and industry, with applications including real estate, public health, crime mapping, national defense, sustainable development, natural resources, climatology, landscape architecture, archaeology, regional and community planning, transportation and logistics. GIS is also diverging into location-based services, which allows GPS-enabled mobile devices to display their location in relation to fixed objects (nearest restaurant, gas station, fire hydrant) or mobile objects (friends, children, police car), or to relay their position back to a central server for display or other processing.

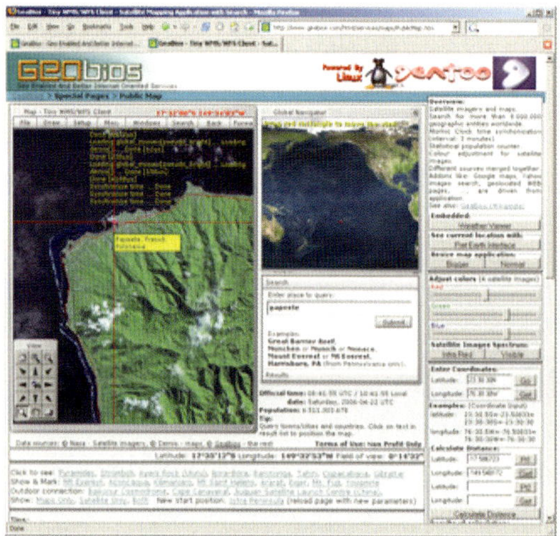

GeaBios – tiny WMS/WFS client (Flash/DHTML)

Open Geospatial Consortium Standards

The Open Geospatial Consortium (OGC) is an international industry consortium of 384 companies, government agencies, universities, and individuals participating in a consensus process to develop publicly available geoprocessing specifications. Open interfaces and protocols defined by OpenGIS Specifications support interoperable solutions that "geo-enable" the Web, wireless and location-based services, and mainstream IT, and empower technology developers to make complex spatial information and services accessible and useful with all kinds of applications. Open Geospatial Consortium protocols include Web Map Service, and Web Feature Service.

GIS products are broken down by the OGC into two categories, based on how completely and accurately the software follows the OGC specifications.

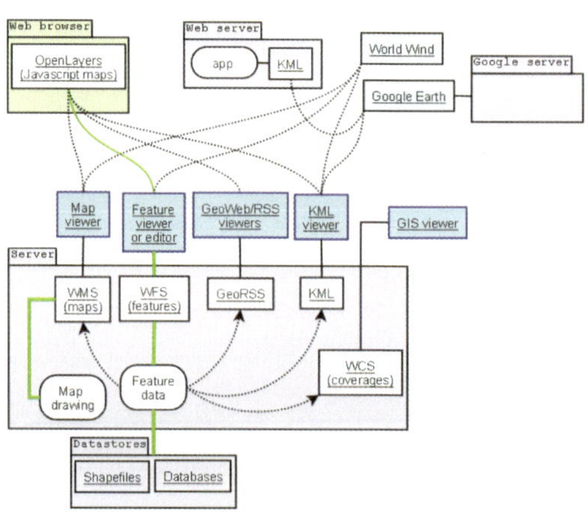

OGC standards help GIS tools communicate.

Compliant Products are software products that comply to OGC's OpenGIS Specifications. When a product has been tested and certified as compliant through the OGC Testing Program, the product is automatically registered as "compliant" on this site.

Implementing Products are software products that implement OpenGIS Specifications but have not yet passed a compliance test. Compliance tests are not available for all specifications. Developers can register their products as implementing draft or approved specifications, though OGC reserves the right to review and verify each entry.

Web Mapping

In recent years there has been a proliferation of free-to-use and easily accessible mapping software such as the proprietary web applications Google Maps and Bing Maps, as well as the free and open-source alternative OpenStreetMap. These services give the public access to huge amounts of geographic data; perceived by many users to be as trustworthy and usable as professional information.

Some of them, like Google Maps and OpenLayers, expose an application programming interface (API) that enable users to create custom applications. These toolkits commonly offer street maps, aerial/satellite imagery, geocoding, searches, and routing functionality. Web mapping has also uncovered the potential of crowdsourcing geodata in projects like OpenStreetMap, which is a collaborative project to create a free editable map of the world. These mashupprojects have been proven to provide a high level of value and benefit to end users outside that possible through traditional geographic information.

Adding the Dimension of Time

The condition of the Earth's surface, atmosphere, and subsurface can be examined by feeding satellite data into a GIS. GIS technology gives researchers the ability to examine the variations in Earth processes over days, months, and years. As an example, the changes in vegetation vigor through a growing season can be animated to determine when drought was most extensive in a particular region. The resulting graphic represents a rough measure of plant health. Working with two variables over time would then allow researchers to detect regional differences in the lag between a decline in rainfall and its effect on vegetation.

GIS technology and the availability of digital data on regional and global scales enable such analysis. The satellite sensor output used to generate a vegetation graphic is produced for example by the advanced very-high-resolution radiometer(AVHRR). This sensor system detects the amounts of energy reflected from the Earth's surface across various bands of the spectrum for surface areas of about 1 square kilometer. The satellite sensor produces images of a particular location on the Earth twice a day. AVHRR and more recently the moderate-resolution imaging spectroradiometer (MODIS) are only two of many sensor systems used for Earth surface analysis.

In addition to the integration of time in environmental studies, GIS is also being explored for its ability to track and model the progress of humans throughout their daily routines. A concrete example of progress in this area is the recent release of time-specific population data by the U.S. Census. In this data set, the populations of cities are shown for daytime and evening hours highlighting the pattern of concentration and dispersion generated by North American commuting patterns. The manipulation and generation of data required to produce this data would not have been possible without GIS.

Using models to project the data held by a GIS forward in time have enabled planners to test policy decisions using spatial decision support systems.

Semantics

Tools and technologies emerging from the World Wide Web Consortium's Semantic Web are proving useful for data integration problems in information systems. Correspondingly, such technologies have been proposed as a means to facilitate interoperability and data reuse among GIS applications. and also to enable new analysis mechanisms.

Ontologies are a key component of this semantic approach as they allow a formal, machine-readable specification of the concepts and relationships in a given domain. This in turn allows a GIS to focus on the intended meaning of data rather than its syntax or structure. For example, reasoning that a land cover type classified as *deciduous needle leaf trees* in one dataset is a specialization or subset of land cover type *forest* in another more roughly classified dataset can help a GIS automatically merge the two datasets under the more general land cover classification. Tentative ontologies have been developed in areas related to GIS applications, for example the hydrology ontology developed by the Ordnance Survey in the United Kingdom and the SWEET ontologiesdeveloped by NASA's Jet Propulsion Laboratory. Also, simpler ontologies and semantic metadata standards are being proposed by the W3C Geo Incubator Group to represent geospatial data on the web. GeoSPARQL is a standard developed by the Ordnance Survey, United States Geological Survey, Natural Resources Canada, Australia's Commonwealth Scientific and Industrial Research Organisation and others to support ontology creation and reasoning using well-understood OGC literals (GML, WKT), topological relationships (Simple Features, RCC8, DE-9IM), RDF and the SPARQL database query protocols.

Recent research results in this area can be seen in the International Conference on Geospatial Semantics and the Terra Cognita – Directions to the Geospatial Semantic Web workshop at the International Semantic Web Conference.

Implications of GIS in Society

With the popularization of GIS in decision making, scholars have begun to scrutinize the social and political implications of GIS. GIS can also be misused to distort reality for individual and political gain. It has been argued that the production, distribution, utilization, and representation of geographic information are largely related with the social context and has the potential to increase citizen trust in government. Other related topics include discussion on copyright, privacy, and censorship. A more optimistic social approach to GIS adoption is to use it as a tool for public participation.

GIS in Education

At the end of the 20th century, GIS began to be recognized as tools that could be used in the classroom. The benefits of GIS in education seem focused on developing spatial thinking, but there is not enough bibliography or statistical data to show the concrete scope of the use of GIS in education around the world, although the expansion has been faster in those countries where the curriculum mentions them.

GIS seem to provide many advantages in teaching geography because they allow for analysis based on real geographic data and also help raise many research questions from teachers and students in classrooms, as well as they contribute to improvement in learning by developing spatial and geographical thinking and, in many cases, student motivation.

GIS in Local Government

GIS is proven as an organization-wide, enterprise and enduring technology that continues to change how local government operates. Government agencies have adopted GIS technology as a method to better manage the following areas of government organization:

- Economic Development departments use interactive GIS mapping tools, aggregated with other data (demographics, labor force, business, industry, talent) along with a database of available commercial sites and buildings in order to attract investment and support existing business. Businesses making location decisions can use the tools to choose communities and sites that best match their criteria for success. GIS Planning's ZoomProspector Enterprise an Intelligence Components software is the industry leader, servicing more than 60% of the US population, more than 30% of Canadians, and locations in the UK and Switzerland. You can see an example of these tools here on the state of Pennsylvania's Department of Community and Economic Development website.
- Public Safety operations such as Emergency Operations Centers, Fire Prevention, Police and Sheriff mobile technology and dispatch, and mapping weather risks.
- Parks and Recreation departments and their functions in asset inventory, land conservation, land management, and cemetery management.
- Public Works and Utilities, tracking water and stormwater drainage, electrical assets, engineering projects, and public transportation assets and trends.
- Fiber Network Management for interdepartmental network assets
- School analytical and demographic data, asset management, and improvement/expansion planning
- Public Administration for election data, property records, and zoning/management.

The Open Data initiative is pushing local government to take advantage of technology such as GIS technology, as it encompasses the requirements to fit the Open Data/Open Government model of transparency. With Open Data, local government organizations can implement Citizen Engagement applications and online portals, allowing citizens to see land information, report potholes and signage issues, view and sort parks by assets, view real-time crime rates and utility repairs, and much more. The push for open data within government organizations is driving the growth in local government GIS technology spending, and database management.

Four Dimensional GIS

The fourth dimension in Geographic Information Systems (GIS) is time. The first two dimensions are what make up a 2D paper map, specifically a grid or coordinate system on which geographic features are rendered. The third dimension is the presentation of height or elevations, which allows for features on a map to be represented in 3D. The fourth dimension accounts for change in mapped features over time.

In the past, most GIS data was used primarily for planning, covered large areas, and was updated in-frequently, perhaps annually. Even with GIS data that is updated more frequently, like Landsat satellite imagery of earth (taken every 16 days over the past 40 years), most change analysis in the past was done on an annualized basis between two adjoining years, and visualized as a third data-set. Viewing time aware data in a series, like in a video, was limited to GIS desktop applications. Consequently, time aware visualization was not widely done in GIS. That all changed when ESRI released version 10.1 of their desktop and server GIS software. Streaming large amounts of data became possible, opening the way to a high quality experience with time aware data on the web. This, in conjunction with an explosion of services providing real-time spatial data, like the national weather service and Twitter, means the time has come for the fourth dimension in GIS.

Capitalizing on the success of the Osprey Dashboard Public, the Center for GIS (CGIS) is currently building an internal dashboard application for the Maryland Emergency Management Agency (MEMA) that takes advantage of time aware data. Having the ability to look back in time at the real-time feeds that MEMA aggregates for the Osprey Dashboard will allow emergency managers to compare a current event to past events and assist with planning, exercises, and after action analysis.

Multimedia or Hypermedia GIS

Multimedia/hypermedia GIS allow the user to access a wide range of georeferenced multimedia data (e.g., simulations, sounds and videos) by selecting resources from a georeferenced image map base. A map serving as the primary index to multimedia data in a multimedia geo-representation is termed a *hypermap*. Multimedia and virtual geo-representations can be stored either in extended relational databases, object databases or in application-specific data stores. Raper (2000) mentions several examples including a multimedia atlas that has been developed for the tourist information system in China.

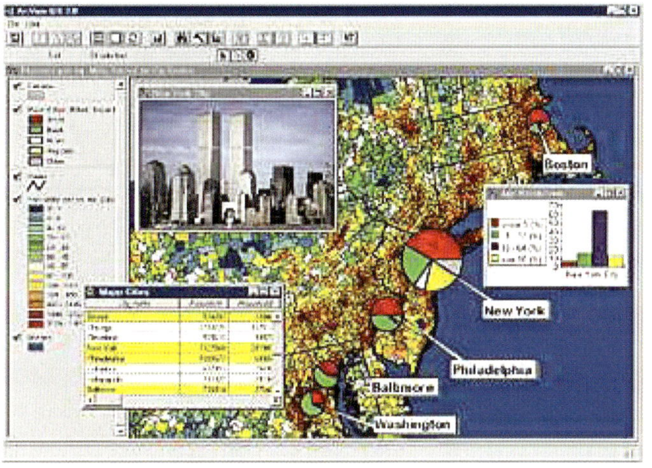

Imagine having a real-time (using telecommunication links) multimedia GIS that can audio-visually monitor different locations at the same time; one application of such system could be as a decision support tool to improve the analysis of traffic problems and the assessment of noise in urban planning and management.

Since video is a spatio-temporal projection of the world in imagery, it can be considered capable of fully multi-dimensional geo-representation. It can be used for mapping from moving platforms, for target positioning, for the measurement of geo-phenomena such as air pollution, and as a process monitoring system. Video imaging has also become a recognised and valuable new technique of remote sensing since the 1990s.

Web GIS

Web GIS is a technology that is used to display and analyze spatial data on the Internet. It combines the advantages of both the Internet and GIS. It offers public a new means to access spatial information without owning expensive GIS software.

Web GIS as known standing for "Geographical Information Systems" could be a little too technical, in simple terms can be summarized to mean a type of distributed information system which came to take over the traditional short-distanced desktop GIS. The essential components it comprises are "the server" and "the client" where the server is the GIS server and the client being web browser or even an application on the mobile phone. Therefore, it's a way of enabling communication between the server and the client.

Here is how Web GIS works over a connection:

- The client reaches the server through a URL which is specific to the server concerned
- Requests from the clients to the server are sent relying on the HTTP (Hypertext Transfer Protocol) standardization
- The server then responds to the requests and send them back through HTTPi.e. The

various formats the response from the server could be in are HTML (Hypertext Mark-up Language); the common form, XML (Extensible Mark-up Language), binary image or JSON (JavaScript Object Notation).

1. Web GIS Allows a Large Number of users

When you have a large connection on clients then WebGIS would be the right way to go for communication as it allows a large number of users from anywhere to have access to the server all at the same time. The number of clients accessible spontaneously depends on how high the performance is as well as its scalability unlike the traditional desktop GIS where only one person is allowed at a time. The higher the performance, the higher the accessibility limit spontaneously.

2. Web GIS can be useful when Limited in Budget

Screening at the relativity between the number of users and the budget required, shows that the budget in installing Web GIS is not as high as traditional desktop GIS. This can be seen through accessibility of content from the internet. It is similar for Web GIS where one doesn't have to pay to have access to Web GIS except for the provider who also still minimizes cost through Web GIS as they are installing only one system and avoid getting every client their PCs. So very useful for large number of clients, remotely accessed, but also very economical way of communication across.

3. Web GIS is useful if the Network is over the Globe

HTTP is very trusted and allowed in almost every network in the world, therefore all organization open their firewalls to accept requests and send responses through local networks. This is of a big advantage as the nature of Web GIS is from HTTP, meaning Web GIS can be used to introduce applications accessible in PCs or mobile phones in almost everywhere in the world. Therefore, this will make communication over the global network possible.

4. Web GIS is Ideal when the Network has Background-diverted Clients

One of the very good properties of Web GIS is that it is easy to use and doesn't require one to be a professional or an expert in GIS as in desktop GIS where it was intended for professional users only. Web GIS is as easy as using a normal website as it is mainly designed to improve usage simplicity, usage coverage and convenience. This had been a success making Web GIS much easier network communicating facilitator.

Web GIS is very useful and very widely used especially for mobile applications from maps, locators, tagging personal photos, displaying Wi-Fi hotspots, linking information and data base between branches of the same company across long distance and many more ways.

Virtual Reality GIS

Virtual reality (VR) has been integrated with GIS for some time, with academic and industry development ongoing since the 1990s. What is different now is data integration and powerful platforms no longer require specialized workstations or high performance computing. From Esri's CityEngine to more specialized software such as GeoScope, VR, or what has been called VRGIS, has now moved into area such as utility in Big Data applications, VR within mobile devices, and allowing users to more easily create sophisticated VR visualizations without programming.

Using Virtual Reality and GIS in Urban Planning

One area of focus is in creating VR applications that provide realistic visualizations of traffic data, where its utility for traffic management is noted. The integration of real-time data and VR allows traffic management to more easily look at varied layers of transport, while also gaining ability to forecast patterns based on historical or expected patterns. Applying it within web technologies also allows easier sharing of data, where managers and those using transport can readily access needed data. Applications such as XEarth have incorporated 3D building model data, residents information, real-time and historical traffic data into their platform, allowing for a powerful understanding of city data where everything from current traffic patterns to up-to-date residential information, where the data are applied within a human-computer interaction framework.

Applications with complex data demand have become a major challenge for VRGIS, particularly as these application have incorporated spatial query functions so that desired multi-dimension information can be found more easily, allowing users to better take advantage of analytical capabilities within GIS while still using VR visualization. Addressing this challenge has included applying multi-task parallel scheduling algorithms that can more quickly process complex data and not just simply look at high volume data. Such approaches allow faster processing of data, allowing for visual and query information to appear more quickly even in more typical desktop environments. In effect, data has become complex in structure and not just in volume, where VR platforms are particularly well made for visualizing such complex information in their appropriate dimensions.

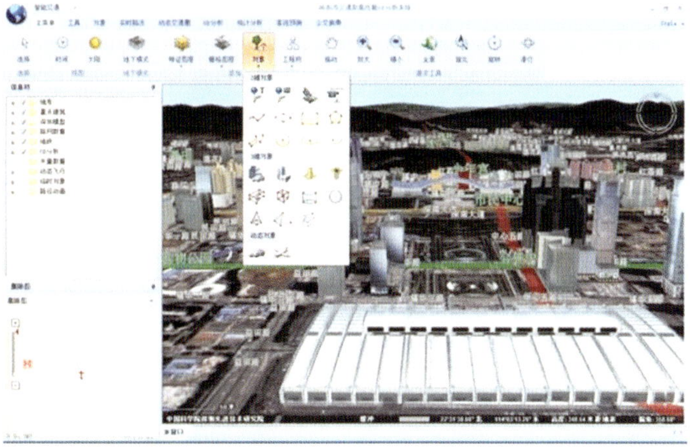

VRGIS INTERFACE. SCREENSHOT: WANG ET. AL, 2017.

VRGIS and Teaching

While many VRGIS platforms and applications have focused on urban data, other applications of VRGIS have been in the area of education, including in the teaching of geography, where the integration of HMD glasses allow users to create immersive and interactive experiences while conduction complex spatial analysis. The user, in this case, can apply touch-less interaction to manipulate a given 3D scene. Other applications of VRGIS have been to understand complex data. In climate modeling, where numerous variables make understanding weather phenomena for given regions an often very complex tax, VRGIS platforms have been considered a way to more easily to visualize complex scientific data in given spatial regions. In applications, the layering of data in a 3D environment allows users to query and look at other relevant information that may indicate a clearer understanding of factors affecting climate.

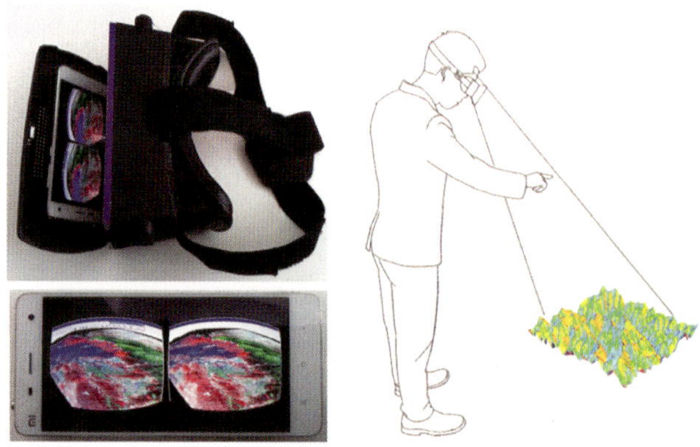

Virtual reality glasses provides a touch-less interaction to manipulate a given 3d scene

While VRGIS has made large strides in the past few years, major challenges remain. First, the integration of VR or even augmented reality with mobile devices continues to be a major focus. The current challenge is to make light and small technologies where VR applications can still be utilized to their full extent while access live and large amounts of data. While there has been progress in this area, bandwidth and data limitations often mean that mobile devices cannot easily display complex, 3D data visuals as easily as many desktop platforms.

References

- Haque, Akhlaque (2015). Surveillance, Transparency and Democracy: Public Administration in the Information Age. Tuscaloosa, AL: University of Alabama Press. pp. 70–73. ISBN 978-0817318772
- Chang, K. T. (1989). "A comparison of techniques for calculating gradient and aspect from a gridded digital elevation model". International Journal of Geographical Information Science. 3 (4): 323–334. doi:10.1080/02693798908941519
- Tomlin, C. Dana (1990). Geographic information systems and cartographic modeling. Prentice Hall series in geographic information science. Prentice Hall. Retrieved 2017-01-05
- Ma, Y.; Guo, Y.; Tian, X.; Ghanem, M. (2011). "Distributed Clustering-Based Aggregation Algorithm for Spatial Correlated Sensor Networks" (PDF). IEEE Sensors Journal. 11 (3): 641. Bibcode:2011ISenJ..11..641M. doi:10.1109/JSEN.2010.2056916
- Sinton, Diana Stuart; Lund, Jennifer J., eds. (2007). Understanding place: GIS and mapping across the curriculum. Redlands, CA: ESRI Press. ISBN 9781589481497. OCLC 70866933
- Jones, K.H. (1998). "A comparison of algorithms used to compute hill slope as a property of the DEM". Computers and Geosciences. 24 (4): 315–323. Bibcode:1998CG.....24..315J. doi:10.1016/S0098-3004(98)00032-6
- Marwick, Ben; Hiscock, Peter; Sullivan, Marjorie; Hughes, Philip (July 2017). "Landform boundary effects on Holocene forager landscape use in arid South Australia". Journal of Archaeological Science: Reports. doi:10.1016/j.jasrep.2017.07.004

2

Remote Sensing: Techniques and Applications

Remote sensing enables the acquisition of information about an event, object or phenomenon without making any direct contact with the area or object under observation. This chapter includes valuable insights into significant topics like water remote sensing, multispectral imaging, normalized difference vegetation index and GIS in environmental contamination, which are crucial for a comprehensive understanding of techniques and applications of GIS.

We perceive the surrounding world through our five senses. Some senses (touch and taste) require contact of our sensing organs with the objects. However, we acquire much information about our surrounding through the senses of sight and hearing which do not require close contact between the sensing organs and the external objects. In another word, we are performing Remote Sensing all the time.

Generally, Remote sensing refers to the activities of recording/observing/perceiving (sensing) objects or events at far away (remote) places. In remote sensing, the sensors are not in direct contact with the objects or events being observed. The information needs a physical carrier to travel from the objects/events to the sensors through an intervening medium. The electromagnetic radiation is normally used as an information carrier in remote sensing. The output of a remote sensing system is usually an image representing the scene being observed. A further step of image analysis and interpretation is required in order to extract useful information from the image. The human visual system is an example of a remote sensing system in this general sense.

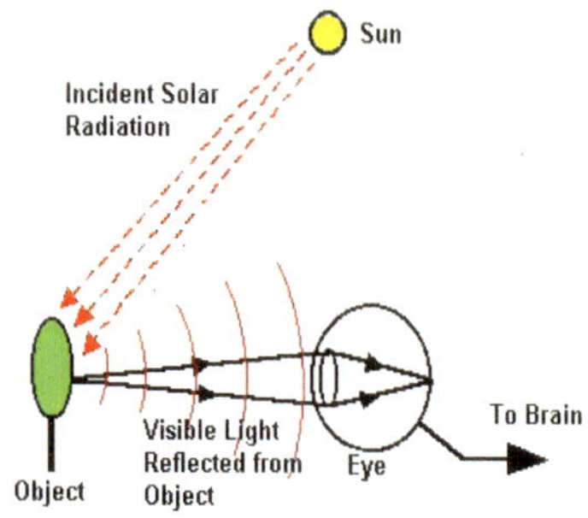

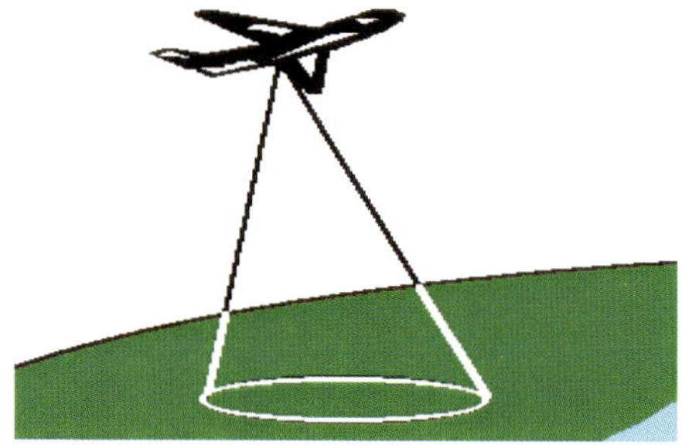

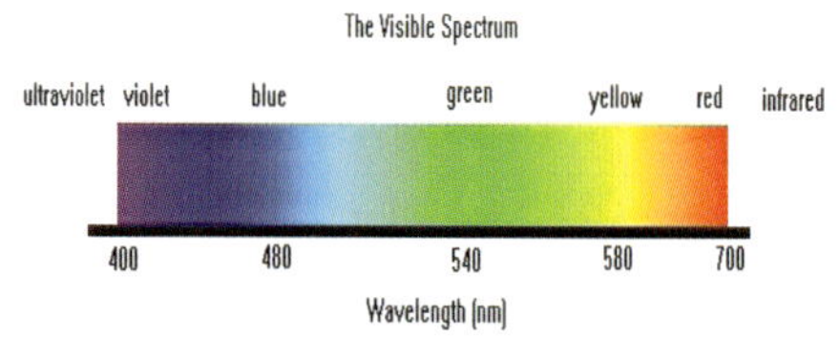

In a more restricted sense, remote sensing usually refers to the technology of acquiring information about the earth's surface (land and ocean) and atmosphere using sensors onboard airborne(aircraft, balloons) or spaceborne (satellites, space shuttles) platforms.

Special cameras collect remotely sensed images of the Earth, which help researchers "sense" things about the Earth. Some examples are:

- Cameras on satellites and airplanes take images of large areas on the Earth's surface, allowing us to see much more than we can standing on the ground.
- Sonar systems on ships can be used to create images of the ocean floor without needing to travel to the bottom of the ocean.
- Cameras on satellites can be used to make images of temperature changes in the oceans.
- Some specific uses of remotely sensed images of the Earth include:
  - ◇ Large forest fires can be mapped from space, allowing rangers to see a much larger area than from the ground.
  - ◇ Tracking clouds to help predict the weather or watch erupting volcanos, and help watch for dust storms.
  - ◇ Tracking the growth of a city and changes in farmland or forests over several years or even decades.
  - ◇ Mapping the ocean bottom - Discovery and mapping of the rugged topography of the ocean floor (e.g., huge mountain ranges, deep canyons, and the "magnetic striping" on the ocean floor).

Passive sensors gather radiation that is emitted or reflected by the object or surrounding areas. Reflected sunlight is the most common source of radiation measured by passive sensors. Examples of passive remote sensors include film photography, infrared, charge-coupled devices, and radiometers. Active collection, on the other hand, emits energy in order to scan objects and areas whereupon a sensor then detects and measures the radiation that is reflected or backscattered from the target. RADAR and LiDAR are examples of active remote sensing where the time delay between emission and return is measured, establishing the location, speed and direction of an object.

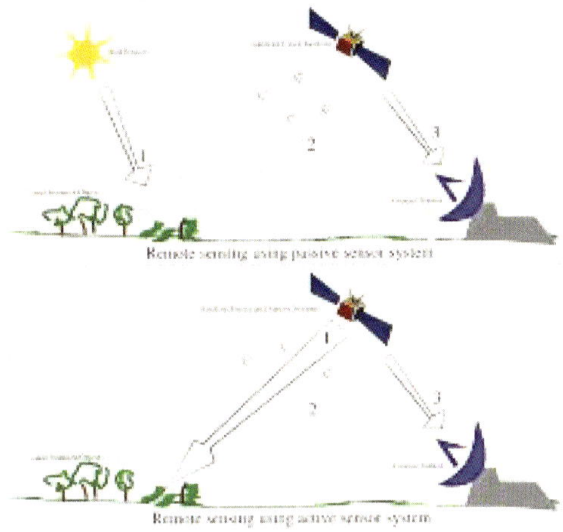

Illustration of Remote Sensing

Remote sensing makes it possible to collect data of dangerous or inaccessible areas. Remote sensing applications include monitoring deforestation in areas such as the Amazon Basin, glacial features in Arctic and Antarctic regions, and depth sounding of coastal and ocean depths. Military collection during the Cold War made use of stand-off collection of data about dangerous border areas. Remote sensing also replaces costly and slow data collection on the ground, ensuring in the process that areas or objects are not disturbed.

Orbital platforms collect and transmit data from different parts of the electromagnetic spectrum, which in conjunction with larger scale aerial or ground-based sensing and analysis, provides researchers with enough information to monitor trends such as El Niño and other natural long and short term phenomena. Other uses include different areas of the earth sciences such as natural resource management, agricultural fields such as land usage and conservation, and national security and overhead, ground-based and stand-off collection on border areas.

Types of Data Acquisition Techniques

The basis for multispectral collection and analysis is that of examined areas or objects that reflect or emit radiation that stand out from surrounding areas. For a summary of major remote sensing satellite systems see the overview table.

Applications of Remote Sensing

- Conventional radar is mostly associated with aerial traffic control, early warning, and certain large scale meteorological data. Doppler radar is used by local law enforcements' monitoring of speed limits and in enhanced meteorological collection such as wind speed and direction within weather systems in addition to precipitation location and intensity. Other types of active collection includes plasmas in the ionosphere. Interferometric synthetic aperture radar is used to produce precise digital elevation models of large scale terrain.

- Laser and radar altimeters on satellites have provided a wide range of data. By measuring the bulges of water caused by gravity, they map features on the seafloor to a resolution of a mile or so. By measuring the height and wavelength of ocean waves, the altimeters measure wind speeds and direction, and surface ocean currents and directions.
- Ultrasound (acoustic) and radar tide gauges measure sea level, tides and wave direction in coastal and offshore tide gauges.
- Light detection and ranging (LIDAR) is well known in examples of weapon ranging, laser illuminated homing of projectiles. LIDAR is used to detect and measure the concentration of various chemicals in the atmosphere, while airborne LIDAR can be used to measure heights of objects and features on the ground more accurately than with radar technology. Vegetation remote sensing is a principal application of LIDAR.
- Radiometers and photometers are the most common instrument in use, collecting reflected and emitted radiation in a wide range of frequencies. The most common are visible and infrared sensors, followed by microwave, gamma ray and rarely, ultraviolet. They may also be used to detect the emission spectra of various chemicals, providing data on chemical concentrations in the atmosphere.
- Stereographic pairs of aerial photographs have often been used to make topographic maps by imagery and terrain analysts in trafficability and highway departments for potential routes, in addition to modelling terrestrial habitat features.
- Simultaneous multi-spectral platforms such as Landsat have been in use since the 1970s. These thematic mappers take images in multiple wavelengths of electro-magnetic radiation (multi-spectral) and are usually found on Earth observation satellites, including the Landsat program or the IKONOS satellite. Maps of land cover and land use from thematic mapping can be used to prospect for minerals, detect or monitor land usage, detect invasive vegetation, deforestation, and examine the health of indigenous plants and crops, including entire farming regions or forests. Landsat images are used by regulatory agencies such as KYDOW to indicate water quality parameters including Secchi depth, chlorophyll a density and total phosphorus content. Weather satellites are used in meteorology and climatology.
- Hyperspectral imaging produces an image where each pixel has full spectral information with imaging narrow spectral bands over a contiguous spectral range. Hyperspectral imagers are used in various applications including mineralogy, biology, defence, and environmental measurements.
- Within the scope of the combat against desertification, remote sensing allows researchers to follow up and monitor risk areas in the long term, to determine desertification factors, to support decision-makers in defining relevant measures of environmental management, and to assess their impacts.

Geodetic

- Geodetic remote sensing can be gravimetric or geometric. Overhead gravity data collection was first used in aerial submarine detection. This data revealed minute perturbations in the Earth's gravitational field that may be used to determine changes in the mass distribution of the Earth, which in turn may be used for geophysical studies, as in GRACE. Geometric remote sensing includes position and deformation imaging using InSAR, LIDAR, etc.

Acoustic and Near-acoustic

- Sonar: *passive sonar*, listening for the sound made by another object (a vessel, a whale etc.); *active sonar*, emitting pulses of sounds and listening for echoes, used for detecting, ranging and measurements of underwater objects and terrain.
- Seismograms taken at different locations can locate and measure earthquakes (after they occur) by comparing the relative intensity and precise timings.
- Ultrasound: Ultrasound sensors, that emit high frequency pulses and listening for echoes, used for detecting water waves and water level, as in tide gauges or for towing tanks.

To coordinate a series of large-scale observations, most sensing systems depend on the following: platform location and the orientation of the sensor. High-end instruments now often use positional information from satellite navigation systems. The rotation and orientation is often provided within a degree or two with electronic compasses. Compasses can measure not just azimuth (i. e. degrees to magnetic north), but also altitude (degrees above the horizon), since the magnetic field curves into the Earth at different angles at different latitudes. More exact orientations require gyroscopic-aided orientation, periodically realigned by different methods including navigation from stars or known benchmarks.

Data Processing

Generally speaking, remote sensing works on the principle of the *inverse problem*. While the object or phenomenon of interest (the state) may not be directly measured, there exists some other variable that can be detected and measured (the observation) which may be related to the object of interest through a calculation. The common analogy given to describe this is trying to determine the type of animal from its footprints. For example, while it is impossible to directly measure temperatures in the upper atmosphere, it is possible to measure the spectral emissions from a known chemical species (such as carbon dioxide) in that region. The frequency of the emissions may then be related via thermodynamics to the temperature in that region.

The quality of remote sensing data consists of its spatial, spectral, radiometric and temporal resolutions.

Spatial Resolution

The size of a pixel that is recorded in a raster image – typically pixels may correspond to square areas ranging in side length from 1 to 1,000 metres (3.3 to 3,280.8 ft).

Spectral Resolution

The wavelength of the different frequency bands recorded – usually, this is related to the number of frequency bands recorded by the platform. Current Landsat collection is that of seven bands, including several in the infrared spectrum, ranging from a spectral resolution of 0.7 to 2.1 μm. The Hyperion sensor on Earth Observing-1 resolves 220 bands from 0.4 to 2.5 μm, with a spectral resolution of 0.10 to 0.11 μm per band.

Radiometric Resolution

The number of different intensities of radiation the sensor is able to distinguish. Typically, this ranges from 8 to 14 bits, corresponding to 256 levels of the gray scale and up to 16,384 intensities or "shades" of colour, in each band. It also depends on the instrument noise.

Temporal Resolution

The frequency of flyovers by the satellite or plane, and is only relevant in time-series studies or those requiring an averaged or mosaic image as in deforesting monitoring. This was first used by the intelligence community where repeated coverage revealed changes in infrastructure, the deployment of units or the modification/introduction of equipment. Cloud cover over a given area or object makes it necessary to repeat the collection of said location.

In order to create sensor-based maps, most remote sensing systems expect to extrapolate sensor data in relation to a reference point including distances between known points on the ground. This depends on the type of sensor used. For example, in conventional photographs, distances are accurate in the center of the image, with the distortion of measurements increasing the farther you get from the center. Another factor is that of the platen against which the film is pressed can cause severe errors when photographs are used to measure ground distances. The step in which this problem is resolved is called georeferencing, and involves computer-aided matching of points in the image (typically 30 or more points per image) which is extrapolated with the use of an established benchmark, "warping" the image to produce accurate spatial data. As of the early 1990s, most satellite images are sold fully georeferenced.

In addition, images may need to be radiometrically and atmospherically corrected.

Radiometric Correction

Allows avoidance of radiometric errors and distortions. The illumination of objects on the Earth surface is uneven because of different properties of the relief. This factor is taken into account in the method of radiometric distortion correction. Radiometric correction gives a scale to the pixel values, e. g. the monochromatic scale of 0 to 255 will be converted to actual radiance values.

Topographic Correction (Terrain Correction)

In rugged mountains, as a result of terrain, the effective illumination of pixels varies considerably. In a remote sensing image, the pixel on the shady slope receives weak illumination and has a low radiance value, in contrast, the pixel on the sunny slope receives strong illumination and has a high radiance value. For the same object, the pixel radiance value on the shady slope will be different from that on the sunny slope. Additionally, different objects may have similar radiance values. These ambiguities seriously affected remote sensing image information extraction accuracy in mountainous areas. It became the main obstacle to further application of remote sensing images. The purpose of topographic correction is to eliminate this effect, recovering the true reflectivity or radiance of objects in horizontal conditions. It is the premise of quantitative remote sensing application.

Atmospheric Correction

Elimination of atmospheric haze by rescaling each frequency band so that its minimum value (usually realised in water bodies) corresponds to a pixel value of 0. The digitizing of data also makes it possible to manipulate the data by changing gray-scale values.

Interpretation is the critical process of making sense of the data. The first application was that of aerial photographic collection which used the following process; spatial measurement through the use of a light table in both conventional single or stereographic coverage, added skills such as the use of photogrammetry, the use of photomosaics, repeat coverage, Making use of objects' known dimensions in order to detect modifications. Image Analysis is the recently developed automated computer-aided application which is in increasing use.

Object-Based Image Analysis (OBIA) is a sub-discipline of GIScience devoted to partitioning remote sensing (RS) imagery into meaningful image-objects, and assessing their characteristics through spatial, spectral and temporal scale.

Old data from remote sensing is often valuable because it may provide the only long-term data for a large extent of geography. At the same time, the data is often complex to interpret, and bulky to store. Modern systems tend to store the data digitally, often with lossless compression. The difficulty with this approach is that the data is fragile, the format may be archaic, and the data may be easy to falsify. One of the best systems for archiving data series is as computer-generated machine-readable ultrafiche, usually in typefonts such as OCR-B, or as digitized half-tone images. Ultrafiche survive well in standard libraries, with lifetimes of several centuries. They can be created, copied, filed and retrieved by automated systems. They are about as compact as archival magnetic media, and yet can be read by human beings with minimal, standardized equipment.

Data Processing Levels

To facilitate the discussion of data processing in practice, several processing "levels" were first defined in 1986 by NASA as part of its Earth Observing System and steadily adopted since then, both internally at NASA and elsewhere these definitions are:

Level	Description
0	Reconstructed, unprocessed instrument and payload data at full resolution, with any and all communications artifacts (e. g., synchronization frames, communications headers, duplicate data) removed.
1a	Reconstructed, unprocessed instrument data at full resolution, time-referenced, and annotated with ancillary information, including radiometric and geometric calibration coefficients and georeferencing parameters (e. g., platform ephemeris) computed and appended but not applied to the Level 0 data (or if applied, in a manner that level 0 is fully recoverable from level 1a data).

1b	Level 1a data that have been processed to sensor units (e. g., radar backscatter cross section, brightness temperature, etc.); not all instruments have Level 1b data; level 0 data is not recoverable from level 1b data.
2	Derived geophysical variables (e. g., ocean wave height, soil moisture, ice concentration) at the same resolution and location as Level 1 source data.
3	Variables mapped on uniform spacetime grid scales, usually with some completeness and consistency (e. g., missing points interpolated, complete regions mosaicked together from multiple orbits, etc.).
4	Model output or results from analysis of lower level data (i. e., variables that were not measured by the instruments but instead are derived from these measurements).

A Level 1 data record is the most fundamental (i. e., highest reversible level) data record that has significant scientific utility, and is the foundation upon which all subsequent data sets are produced. Level 2 is the first level that is directly usable for most scientific applications; its value is much greater than the lower levels. Level 2 data sets tend to be less voluminous than Level 1 data because they have been reduced temporally, spatially, or spectrally. Level 3 data sets are generally smaller than lower level data sets and thus can be dealt with without incurring a great deal of data handling overhead. These data tend to be generally more useful for many applications. The regular spatial and temporal organization of Level 3 datasets makes it feasible to readily combine data from different sources.

While these processing levels are particularly suitable for typical satellite data processing pipelines, other data level vocabularies have been defined and may be appropriate for more heterogeneous workflows.

Remote Sensing Software

Remote sensing data are processed and analyzed with computer software, known as a remote sensing application. A large number of proprietary and open source applications exist to process remote sensing data. Remote sensing software packages include:

- ERDAS IMAGINE from Hexagon Geospatial (Separated from IntergraphSG&I),
- PCI Geomatica
- TNTmips from MicroImages
- IDRISI from Clark Labs
- eCognition from Trimble
- RemoteView made by Overwatch Textron Systems.
- Dragon/ips is one of the oldest remote sensing packages still available, and is in some cases free

Open source remote sensing software includes:

- Opticks (software)
- Orfeo toolbox
- Sentinel Application Platform (SNAP) from the European Space Agency (ESA)

- Others mixing remote sensing and GIS capabilities are: GRASS GIS, ILWIS, QGIS, and TerraLook

According to an NOAA Sponsored Research by Global Marketing Insights, Inc. the most used applications among Asian academic groups involved in remote sensing are as follows: ERDAS 36% (ERDAS IMAGINE 25% & ERMapper 11%); ESRI 30%; ITT Visual Information Solutions ENVI 17%; MapInfo 17%.

Among Western Academic respondents as follows: ESRI 39%, ERDAS IMAGINE 27%, MapInfo 9%, and AutoDesk 7%.

In education, those that want to go beyond simply looking at satellite images print-outs either use general remote sensing software (e.g. QGIS), Google Earth, StoryMaps or a software/ web-app developed specifically for education (e.g. desktop: LeoWorks, online: BLIF).

Water Remote Sensing

Remote Sensing, coupled with Geographic Information System (GIS), is a powerful tool for monitoring of water quality and water pollution. Satellite imageries have been used successfully in determination of various water quality parameters like Total Suspended Solids, turbidity, chlorophyll content, colour, temperature etc. by using the Visible, Reflected Infrared and occasionally Thermal Infrared bands of the Electro Magnetic Spectrum. Remote Sensing techniques have been used in sustainable management of water resources, whichinclude runoff and hydrological modeling, flood management, watershed management, drought management and management of Irrigation Command Areas. Satellite imageries obtained from Landsat Thematic Mapper (TM), Linear Imaging and Self Scanning (LISS) and Wide Field Sensor (WiFS) have been used extensively by various workers for calculation of drainage basin area, drainage density, Normalized Difference Vegetation Index (NDVI) and Leaf Water Content Index (LWCI). Application of Remote Sensing in flood management,drought management and Irrigation Command Area management were demonstrated in India by a number of researchers from time to time. Typical applications include mapping of active flood plains, long-range weather forecasting, estimation of crop acreage and crop production and preparation of "irrigability maps" through land use planning.

Sustainable management of the available water resource is a challenging task for the new millennium. As stated by the World Water Council, "There is a water crisis today. But the crisis is not having too little water to satisfy our needs. It is crisis of managing water so badly that billions of people and the environment – suffer badly" (World Water Council, 2000). Remote Sensing techniques have been used effectively in integrated development and management of water resources of India (Balakrishnan, 1986).

Water has very low spectral reflectance in the visible part of the Electro Magnetic Region (EMR) whereas snow or ice has very high spectral reflectance in visible and near infrared (NIR) part of the EMR. Pure water absorbs nearly all incident energy in both the near infrared and middle infrared (MIR) wavelengths. The low reflectance of water in visible and NIR band has advantage in Remote

Sensing as water becomes clearly distinguishable from either vegetation or soil cover throughout the reflective infrared portion.

Total Radiance (Rt) recorded by a Remote Sensing system over a water body is a function of the electromagnetic energy and is given by the equation:

Rt = Rp + Rs + Rv + Rbwhere, Rp = Atmospheric Path Radiance

Rs = Free-surface Layer Reflectance

Rv = Subsurface Volumetric Reflectance

Rb = Bottom Reflectance

In situ Spectroradiometer measurement of clear water with various levels of clayey and silty soil as suspended sediment shows that the reflectance peak shifts towards longer wavelengths as more suspended sediment is added to the water. Strong chlorophyll a absorption of blue light is observed between wavelengths of 400 and 500 nm and strong chlorophyll a absorption of red light is observed at approximately 675 nm (Lillesand andKiefer, 2000).

Application of visual and digital Remote Sensing techniques and integration of the remotely sensed data in specific layers through the Geographic Information System (GIS) are used by scientists in management of water resources and prediction of natural water related hazards like flood and drought. Visual Remote Sensing has been extensively used in detection of water pollution, lake eutrophication assessment and estimation of flood damage. The technique of visual image interpretation can be used in variety of ways to help monitor water quantity, quality and geographic distribution of water resources.

Function

Water Remote sensing instruments allow to record the color of a water body, which provides information on the presence and abundance of optically active natural water components. The water color spectrum is defined as an apparent optical property (AOP) of the water. This means that the color of the water is influenced by the angular distribution of the light field and by the nature and quantity of the substances in the medium, in this case, water. Thus, the value of this parameter will change with changes in the optical properties and concentrations of the optically active substances in the water, the inherent optical properties or IOPS. The IOPS are independent from the angular distribution of light but they are dependent from the type and substances present in the medium as well. For instance, the diffuse attenuation coefficient of down welling irradiance, Kd (it is often used as an index of water clarity or ocean turbidity) is defined as an AOP, while the absorption coefficient and the scattering coefficient of the medium are defined as IOPS. There are two different approaches to determine the concentration of optically active water components by the study of the spectra. The first approach consist of empirical algorithms based on statistical relationships and the second approach consists of analytical algorithms based on the inversion of calibrated bio-optical models. Accurate calibration of the relationships/models used is an important condition for successful inversion on water remote sensing techniques and the determination of concentration of water quality parameters from observed spectral remote sensing data. Thus, these techniques depend on their ability to record these changes in the spectral signature of light backscattered from water surface and relate these recorded changes to water quality parameters via empirical or analytical approaches. Depending of the water constituents of interest and the sensor used, different parts of the spectrum will be analysed.

Contribution

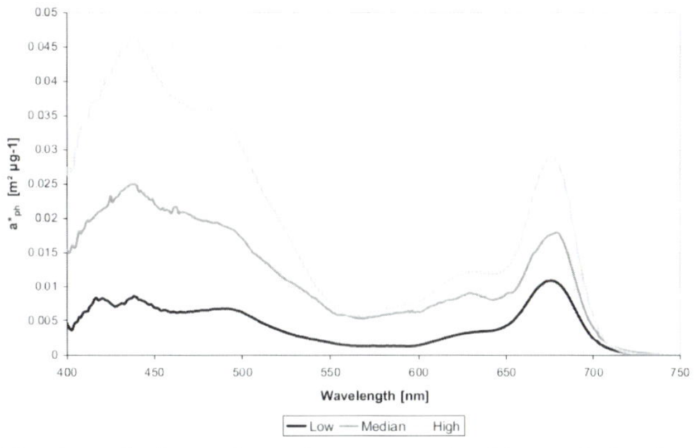

Example of specific phytoplankton absorption spectra. In this graph the characteristic blue and red Ch-a peaks at 438 nm and 676 nm can be seen. Another visible peak is the Cyanophicocianin absorption maximum at 624 nm.

By the use of Optical close range devices (e.g. spectrometers, radiometers), airplanes or helicopters (airborne remote sensing) and satellites (space born remote sensing), the light reflected from the water bodies is measured. For instance, algorithms are used to retrieve parameters such as chlorophyll-a(Chl-a) and Suspended Particulate Matter (SPM) concentration, the absorption by colored dissolved organic matter at 440 nm (aCDOM) and secchi depth. The measurement of these values will give an idea about the water quality of the water body being studied. A very high concentration of green pigments like chlorophyll might indicate the presence of an algal bloom, for example, due to eutrophication processes. Thus, the chlorophyll concentration could be used as a proxy or indicator for the trophic condition of a water body. In the same manner, other optical quality parameters such as suspended particles or Suspended Particulate matter (SPM), Colored Dissolved Organic Matter (CDOM), Transparency (Kd), and chlorophyll-a (Chl-a) can be used to monitor water quality.

Remote Sensing and Water Quality

Spectral properties of water vary with wavelength of incident radiation not only due to the molecular structure, but also due to impurities present in the water body. Hence, Remote Sensing for mapping or monitoring water quality becomes quite complex. The water surface behaves as a partially diffused and partially specular reflector. Specular reflection is uniform at all wavelengths, but absorption and back scatter produce distinctive spectral signature or spectral response pattern. Solar energy that is not sparsly reflected is reflected downward into the water body. This refracted energy is either absorbed or get scattered. The remaining signal is indicative of water quality, which is volumetric reflectance or back scattered energy caused by the material in water. In case of deep water, where the bottom reflectance is negligible, the reflectance comes from the surface of water body. However, for shallow water, the total reflectance is a function of both surface and bottom reflectance.

Spectral signature (or Spectral Response Pattern) of water is dependent on a number of factors viz. time of the year, sun elevation angle, aerosol and molecular content of atmosphere, water vapour content of the atmosphere, specular reflection of skylight from water surface, roughness of water surface, water colour and content of dissolved coloured material in water, characteristics of water surface (film, foam, debris or floating plant), reflectance and absorbance characteristics of suspended particles, multiple reflections and scattering of solar energy in water, depth of water and reflectance of bottom sediment, submerged or emerged vegetation and turbidity of water.

Physical characteristics of water quality can be ascertained through satellite imagery (Moore, 1977). Visible and infrared (reflected) parts of the EMR are most favourable for monitoring of water quality. Thermal Infrared (TIR) is also used occasionally for measuring water quality parameters but the method uses a direct measurement of emitted energy. Microwave region of the EMR is not very useful for water quality assessment due to its limited penetration depth. However, it is useful for detecting oil slicks, oil spills or other surface contamination.

In order to monitor water quality through Remote Sensing, the relationship between water quality parameters and spectral reflectance must be determined. As the reflectance changes with the modified value of water quality parameters, an empirical formula may be used. However, the formula may not be valid in multi temporal domain as the type of constituents in water may not remain constant. Sun elevation angle and atmospheric composition change with time and will affect the relationship between water quality parameter and spectral reflectance.

Multispectral mapping of seagrass has been done where bottom details in clear, calm ocean water have been mapped with penetration of about 20 m between the wavelengths of 0.48 to 0.60 μm using Landsat TM data (Smith and Jensen, 1998). A limitation of this method is that although blue wavelengths have the maximum penetration, extensive scattering occurs resulting in an "underwater haze". Penetration of only a few metres was obtained using the red wavelength.

Multispectral Image

Most satellites today measure energy at many wavelengths. This is called multispectral imaging. Images taken at different wavelengths can be combined to make composite images by displaying the image for each wavelength as red, green, or blue in the final image. These composite images result in color patterns that can be used to identify surface features. Multi-spectral imaging can capture light from frequencies beyond the visible light range, such as infrared. This can allow extraction of additional information that the human eye fails to capture with its receptors for red, green and blue.

A multispectral image consists of several bands of data. For visual display, each band of the image may be displayed one band at a time as a grey scale image, or in combination of three bands at a time as a colour composite image. Interpretation of a multispectral colour composite image will require the knowledge of the spectral reflectance signature of the targets in the scene. In this case, the spectral information content of the image is utilized in the interpretation.

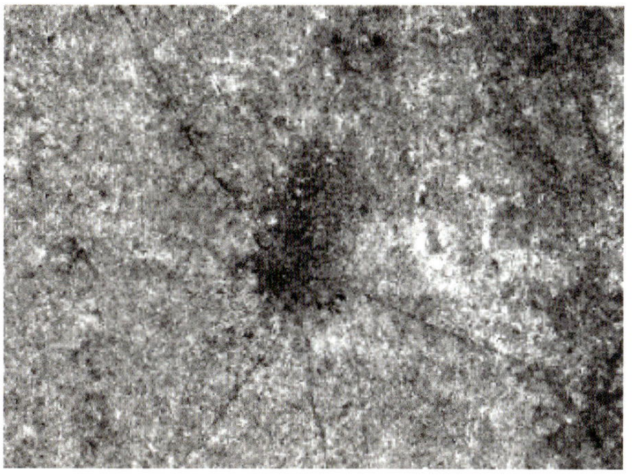

Panchromatic Image (0.52 to 0.90 μm) of Aligarh city and surrounding areas.

The following three images show the three bands of a multispectral image extracted from a LANDSAT multispectral scene at a ground resolution of 28.5 m. The area covered is the same as that shown in the above panchromatic image.

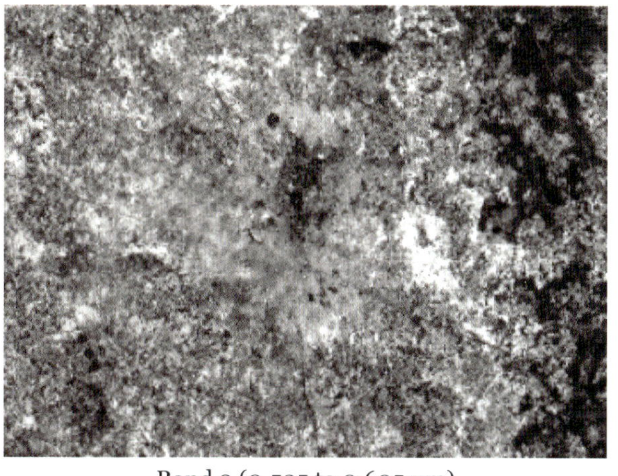

Band 2 (0.525 to 0.605 μm)

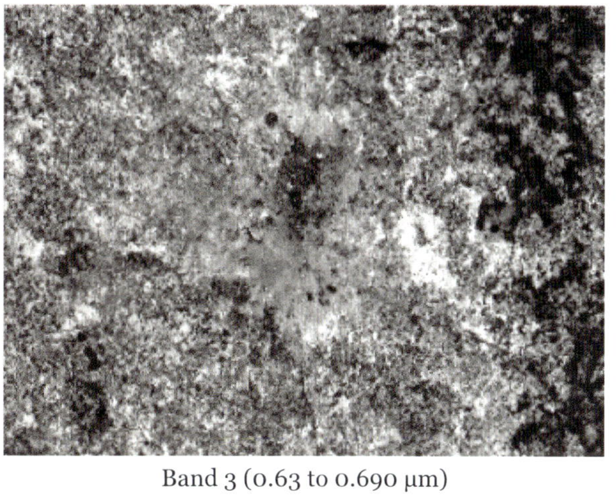

Band 3 (0.63 to 0.690 μm)

Note that both Band 2 (green) and Band 3 (red) bands look almost identical to the panchromatic image shown above.

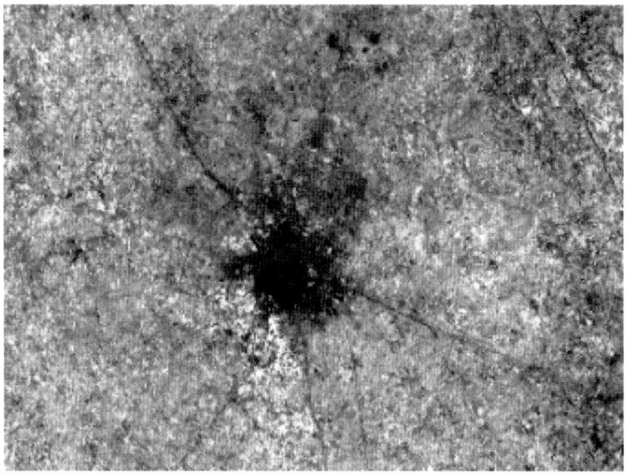

Band 4 (0.75 to 0.90 µm)

In contrast, the vegetated areas now appear bright in the Band 4 (near infrared) band due to high reflectance of leaves in the near infrared wavelength region. Several shades of grey can be identified for the vegetated areas, corresponding to different types of vegetation. Water mass (both the river and the sea) appear dark in the Band 4 (near IR) band.

Multi-spectral images are the main type of images acquired by Remote sensing (RS) radiometers. Multi-spectral is the opposite of panchromatic. Usually satellites have 3 to 7 or more radiometers (Landsat has 7). Each one acquires one digital image (in remote sensing, called a *scene*) in a small band of visible spectra, ranging 0.7 µm to 0.4 µm, called red-green-blue (RGB) region, and going to infra-red wavelengths of 0.7 µm to 10 or more µm, classified as NIR-Near InfraRed, MIR-Middle InfraRed and FIR-Far InfraRed or Thermal. In the Landsat case there are 7 scenes comprising a 7 band multi spectral image.

This technology has also assisted in the interpretation of ancient papyri such as those found at Herculaneum, by imaging the fragments in the infrared range (1000nm). Often the text on the documents appears to be as black ink on black paper to the naked eye. At 1000nm, the difference in light reflectivity makes the text clearly readable.

Applications

Space-based Imaging

Most radiometers for remote sensing (RS) acquire multispectral images. Dividing the spectrum into many bands, multispectral is the opposite of panchromatic, which records only the total intensity of radiation falling on each pixel. Usually, Earth observation satellites have three or more radiometers. Each acquires one digital image (in remote sensing, called a 'scene') in a small spectral band. The bands are grouped into wavelength regions based on the origin of the light and the interests of the researchers.

"Multispectral imaging combines two to five spectral imaging bands of relatively large bandwidth into a single optical system. A multispectral system usually provides a combination of visible (0.4 to 0.7 µm), near infrared (NIR; 0.7 to 1 µm), short-wave infrared (SWIR; 1 to 1.7 µm), mid-wave infrared (MWIR; 3.5 to 5 µm) or long-wave infrared (LWIR; 8 to 12 µm) bands into a single system".

In the case of Landsat satellites, several different band designations have been used, with as many as 11 bands (Landsat 8) comprising a multispectral image. Spectral imaging with a higher radiometric resolution (involving hundreds or thousands of bands), finer spectral resolution (involving smaller bands), or wider spectral coverage may be called hyperspectral or ultraspectral.

Documents and Artworks

The technology has also assisted in the interpretation of ancient papyri, such as those found at Herculaneum, by imaging the fragments in the infrared range (1000 nm). Often, the text on the documents appears to the naked eye as black ink on black paper. At 1000 nm, the difference in how paper and ink reflect infrared light makes the text clearly readable. It has also been used to image the Archimedes palimpsest by imaging the parchment leaves in bandwidths from 365-870 nm, and then using advanced digital image processing techniques to reveal the undertext with Archimedes' work.

Multispectral imaging can be employed for investigation of paintings and other works of art. The painting is irradiated by ultraviolet, visible and infrared rays and the reflected radiation is recorded in a camera sensitive in this regions of the spectrum. The image can also be registered using the transmitted instead of reflected radiation. In special cases the painting can be irradiated by UV, VIS or IR rays and the fluorescence of pigments or varnishes can be registered.

Multispectral imaging has also been used to examine discolorations and stains on old books and manuscripts. Comparing the "spectral fingerprint" of a stain to the characteristics of known chemical substances can make it possible to identify the stain. This technique has been used to examine medical and alchemical texts, seeking hints about the activities of early chemists and the possible chemical substances they may have used in their experiments. Like a cook spilling flour or vinegar on a cookbook, an early chemist might have left tangible evidence on the pages of the ingredients used to make medicines.

Spectral Bands

The wavelengths are approximate; exact values depend on the particular satellite's instruments:

- Blue, 450-515..520 nm, is used for atmosphere and deep water imaging, and can reach depths up to 150 feet (50 m) in clear water.
- Green, 515..520-590..600 nm, is used for imaging vegetation and deep water structures, up to 90 feet (30 m) in clear water.
- Red, 600..630-680..690 nm, is used for imaging man-made objects, in water up to 30 feet (9 m) deep, soil, and vegetation.
- Near infrared (NIR), 750-900 nm, is used primarily for imaging vegetation.
- Mid-infrared (MIR), 1550-1750 nm, is used for imaging vegetation, soil moisture content, and some forest fires.
- Far-infrared (FIR), 2080-2350 nm, is used for imaging soil, moisture, geological features, silicates, clays, and fires.

- Thermal infrared, 10400-12500 nm, uses emitted instead of reflected radiation to image geological structures, thermal differences in water currents, fires, and for night studies.
- Radar and related technologies are useful for mapping terrain and for detecting various objects.

Spectral band usage

For different purposes, different combinations of spectral bands can be used. They are usually represented with red, green, and blue channels. Mapping of bands to colors depends on the purpose of the image and the personal preferences of the analysts. Thermal infrared is often omitted from consideration due to poor spatial resolution, except for special purposes.

- True-color uses only red, green, and blue channels, mapped to their respective colors. As a plain color photograph, it is good for analyzing man-made objects, and is easy to understand for beginner analysts.
- Green-red-infrared, where the blue channel is replaced with near infrared, is used for vegetation, which is highly reflective in near IR; it then shows as blue. This combination is often used to detect vegetation and camouflage.
- Blue-NIR-MIR, where the blue channel uses visible blue, green uses NIR (so vegetation stays green), and MIR is shown as red. Such images allow the water depth, vegetation coverage, soil moisture content, and the presence of fires to be seen, all in a single image.

Many other combinations are in use. NIR is often shown as red, causing vegetation-covered areas to appear red.

Classification

Unlike other Aerial photographic and satellite image interpretation work, these multispectral images do not make it easy to identify directly the feature type by visual inspection. Hence the remote sensing data has to be classified first, followed by processing by various data enhancement techniques so as to help the user to understand the features that are present in the image.

Such classification is a complex task which involves rigorous validation of the training samples depending on the classification algorithm used. The techniques can be grouped mainly into two types.

- Supervised classification techniques
- Unsupervised classification techniques

Supervised classification makes use of training samples. Training samples are areas on the ground for which there is Ground truth, that is, what is there is known. The spectral signatures of the training areas are used to search for similar signatures in the remaining pixels of the image, and we will classify accordingly. This use of training samples for classification is called supervised classification. Expert knowledge is very important in this method since the selection of the training samples and a biased selection can badly affect the accuracy of classification. Popular techniques include the Maximum likelihood principle and Convolutional neural network. The Maximum likelihood principle calculates the probability of a pixel belonging to a class (i.e. feature) and allots the pixel to its most probable class. Newer Convolutional neural network based methods account for both spatial proximity and entire spectra to determine the most likely class.

In case of unsupervised classification no prior knowledge is required for classifying the features of the image. The natural clustering or grouping of the pixel values, i.e. the gray levels of the pixels, are observed. Then a threshold is defined for adopting the number of classes in the image. The finer the threshold value, the more classes there will be. However, beyond a certain limit the same class will be represented in different classes in the sense that variation in the class is represented. After forming the clusters, ground truth validation is done to identify the class the image pixel belongs to. Thus in this unsupervised classification apriori information about the classes is not required. One of the popular methods in unsupervised classification is k-means clustering.

Multispectral data analysis software

- MicroMSI is endorsed by the NGA.
- Opticks is an open-source remote sensing application.
- Multispec is freeware multispectral analysis software.
- Gerbil is open source multispectral visualization and analysis software.

Normalized Difference Vegetation Index

In an effort to monitor major fluctuations in vegetation and understand how they affect the environment, 20 years ago Earth scientists began using satellite remote sensors to measure and map the density of green vegetation over the Earth. Using NOAA's Advanced Very High Resolution Radiometer (AVHRR), scientists have been collecting images of our planet's surface. By carefully measuring the wavelengths and intensity of visible and near-infrared light reflected by the land surface back up into space, scientists use an algorithm called a "Vegetation Index" to quantify the concentrations of green leaf vegetation around the globe. Then by combining the daily Vegetation Indices into 8-, 16-, or 30-day composites, scientists create detailed maps of the Earth's green vegetation density that identify where plants are thriving and where they are under stress (i.e., due to lack of water).

The Normalized Difference Vegetation Index (NDVI) is an index of plant "greenness" or photosynthetic activity, and is one of the most commonly used vegetation indices. Vegetation indices are based on the observation that different surfaces reflect different types of light differently. Photosynthetically active vegetation, in particular, absorbs most of the red light that hits it while reflecting much of the near infrared light. Vegetation that is dead or stressed reflects more red light and less near infrared light. Likewise, non-vegetated surfaces have a much more even reflectance across the light spectrum.

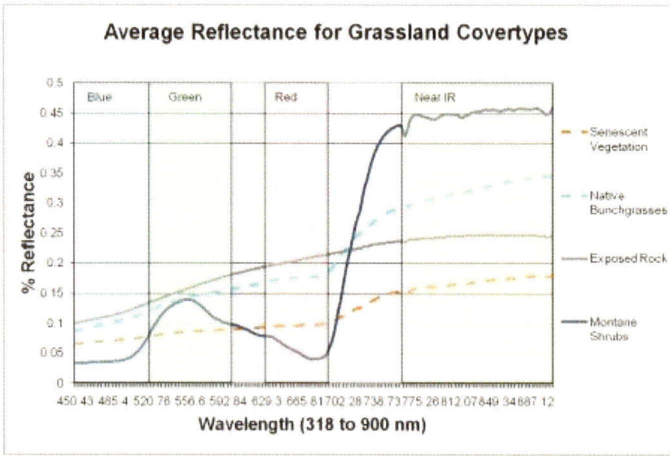

Data for this graph courtesy of the Idaho Chapter of The Nature Conservancy

Reflectance of sunlight from four different land cover types in Hells Canyon, Idaho as measured by a field spectrometer.

GIS in Environmental Contamination

Protection of urban areas requires measurements and observations of a wide range of parameters that can characterize levels of environmental pollution. Mostly, the measurements are focused on air pollution, pollution of surface water and groundwater, soil contamination, noise assessment, waste management and landscape protection. Within these fields, monitoring of environmental pollution is created to estimate short and long term changes, to carry out risk assessment analysis, to develop models for prediction and optimization, and finally, to print maps for the support of decision-making processes. In the past, only heavy industrial areas and large cities were monitored and mapped in the local scale. At present, observations in a global scale are carried out to protect all the environment of our planet. A wide variety of techniques in order to determine individual factors of pollution and stress are used for decision-making processes. A number of automatic and manual-operated stations, chemical analysis in laboratories, satellite technology, risk assessment analysis and modeling systems is applied to improve our knowledge about behavior of various types of pollution and its influence on living organisms. Mapping of pollution together with their sources and the living environment helps to illustrate all the phenomena more transparently. Increasingly more powerful desktop and especially mobile computing systems with their graphic capabilities represent a new possibility for digital mapping. In addition to mapping, advanced analysis can be solved by digital computing systems, which can be represented by the Geographic Information Systems-GIS. There are a number of GISs in the market. Their capabilities are mainly focused on spatial data management that includes a wide range of extensions for spatial analysis, development of three-dimensional map objects, network optimization, image processing, etc. New data inputs represent satellite and aerial images processed by remote sensing, and geopositional systems-GPS. Integration of digital tools for mapping of environmental pollution together with standard monitoring systems and environmental measurements illustrates figure.

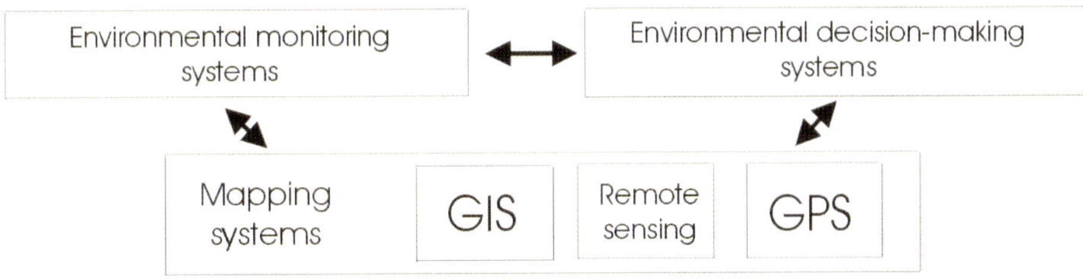

Integration of digital mapping tools in the frame of environmental protection.

Methods

Progress in environmental protection depends on a wide range of components. Besides of data collections and standard decision-making procedures, more sophisticated analysis are needed to effectively solve all the problems. In this part of research, new technology in the frame of the GIS and the mobile computing offers portable solutions. Moreover entering, storing and displaying spatial data in the form of maps, computing capabilities of the mobile PCs together with aerial and satellite images supported by the GPS can be directly used for estimates and analyzes of environmental parameters. The results can be displayed by the GIS in the map themes or transferred via the network. As an example, the GIS for mobile data collections and estimates is illustrated in figure. The schema is complemented by the software tools for development of modules that can modify and extend existing capabilities of the GIS. The selected environmental analysis, which are described by mathematical rules, are implemented in the software extensions of the GIS. There are a few ways of building extended modules in the GISs. The example illustrates the development of the software extensions by the ESRI's ArcPad Application Builder, which represents the software development kit for programming and design. The ArcPad by itself represents the mobile GIS that can run in the standard Windows or Windows CE environment. In addition to entering data and running of the software applications, direct inputs from the GPS and other mobile devices enable online data collection. All the collected data and information in the GIS project can be visualized in the map layers and transferred via the network to the relational database.

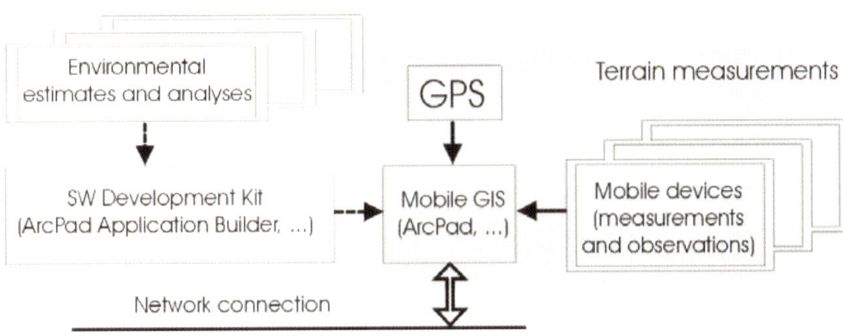

The development and use of environmental analysis in the frame of the GIS mapping.

The methods implemented in the GIS for the support of mapping of environmental pollution are mostly focused on data collection and basic analysis, modeling and predictions. A number of applications, which can be also partly implemented in the mobile GIS, is described in the book: GIS and Environmental Modeling by Goodchild et al. (3). GIS research in the field of water management describes Maidment (6). More general principles of GIS data management and statistical analysis are introduced by Chilès and Delfiner (2), Legendre (4) and Zar (9). GIS mapping and its applications in environmental studies are partly described in the book written by Longley (5). The selection of appropriate methods for the support of the GIS mobile computing depends on the potentiality of development tools, the capacity of storage devices and the power of a computer. The GIS project described in the frame of this paper is focused on applications in the environment of the Pocket PC, which offer high mobility and adequately performance on its level of computing. The map layers for mapping of environmental pollution illustrates figure. The layers are divided into a few classes, which are focused on a specific type of environmental pollution. In addition to the location of pollution in the vector formats (point, line or polygon groups of objects) or raster grids, each object of the layer can contain attributes (the level of pollution, data and time of the location and other specific information). All the special and temporal data can by repeatedly analyzed and preprocessed into new layers.

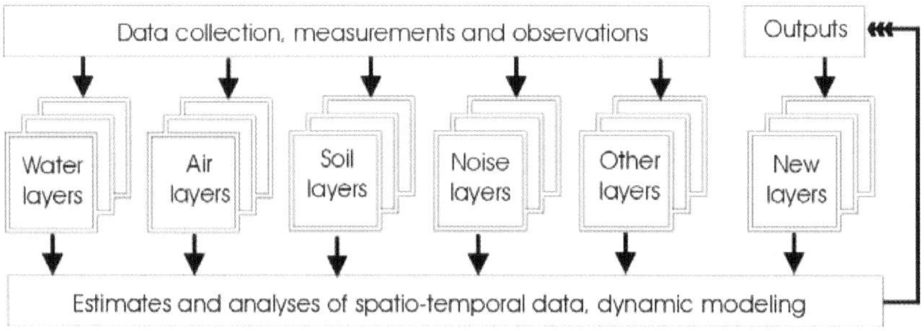

Mapping of environmental pollution by the GIS

In spite of that the estimates and analysis of spatial and temporal data depend on a specific type of pollution, the common tools for the complex evaluation in the frame of the GIS can represent the raster algebra (evaluation of data in regular grids) and the spatial analysis of the vector objects (point, line, polygon objects).

References

- Schowengerdt, Robert A. (2007). Remote sensing: models and methods for image processing (3rd ed.). Academic Press. p. 2. ISBN 978-0-12-369407-2
- Mills, J.P.; et al. (1997). “Photogrammetry from Archived Digital Imagery for Seal Monitoring”. The Photogrammetric Record. 15(89): 715–724. doi:10.1111/0031-868X.00080
- Schiebe, F.R., Harrington, Jr., J.A.; Ritchie, J.C. (1992), “Remote sensing of suspended sediments: The Lake Chicot, Arkansas project”, International Journal of Remote Sensing, 13(8):1487–1509
- Coffey, Valerie C. (1 April 2012). “Multispectral Imaging Moves into the Mainstream”. Optics and Photonics News. 23 (4): 18. doi:10.1364/OPN.23.4.000018. Retrieved 14 May 2018
- Liu, Jian Guo & Mason, Philippa J. (2009). Essential Image Processing for GIS and Remote Sensing. Wiley-Blackwell. p. 4. ISBN 978-0-470-51032-2
- Twiss, S.D.; et al. (2001). “Topographic spatial characterisation of grey seal Halichoerus grypus breeding habitat at a sub-seal size spatial grain”. Ecography. 24 (3): 257–266. doi:10.1111/j.1600-0587.2001.tb00198.x
- Avril, Tom (May 14, 2018). “Scans reveal secrets of medieval ‘Harry Potter’ book and medical texts at Penn”. The Philadelphia Inquirer. Retrieved 14 May 2018
- Grigoriev A.N. (2015). “Method of radiometric distortion correction of multispectral data for the earth remote sensing”. Scientific and Technical Journal of Information Technologies, Mechanics and Optics. 15 (4): 595–602

3

Geographic Coordinate Systems

In geometry, the coordinate system that helps in determining and locating every location on Earth, by specifying a unique set of letters, numbers or symbols, is termed as a geographic coordinate system. The coordinates are chosen so as to represent the location's vertical and horizontal position. This chapter closely examines the different geographic coordinate systems such as geodetic datum and the Cartesian coordinate system.

A geographic coordinate system (GCS) uses a three-dimensional spherical surface to define locations on the earth. A GCS is often incorrectly called a datum, but a datum is only one part of a GCS. A GCS includes an angular unit of measure, a prime meridian, and a datum (based on a spheroid).

A point is referenced by its longitude and latitude values. Longitude and latitude are angles measured from the earth's center to a point on the earth's surface. The angles often are measured in degrees (or in grads). The following figure shows the world as a globe with longitude and latitude values.

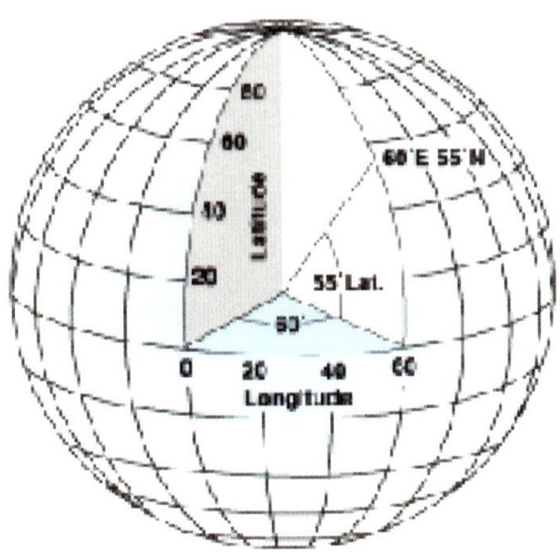

In the spherical system, horizontal lines, or east–west lines, are lines of equal latitude, or parallels. Vertical lines, or north–south lines, are lines of equal longitude, or meridians. These lines encompass the globe and form a gridded network called a graticule.

The line of latitude midway between the poles is called the equator. It defines the line of zero latitude. The line of zero longitude is called the prime meridian. For most geographic coordinate systems, the prime meridian is the longitude that passes through Greenwich, England. Other countries use longitude lines that pass through Bern, Bogota, and Paris as prime meridians. The origin of the graticule (0,0) is defined by where the equator and prime meridian intersect. The globe is then divided into four geographical quadrants that are based on compass bearings from the origin. North and south are above and below the equator, and west and east are to the left and right of the prime meridian.

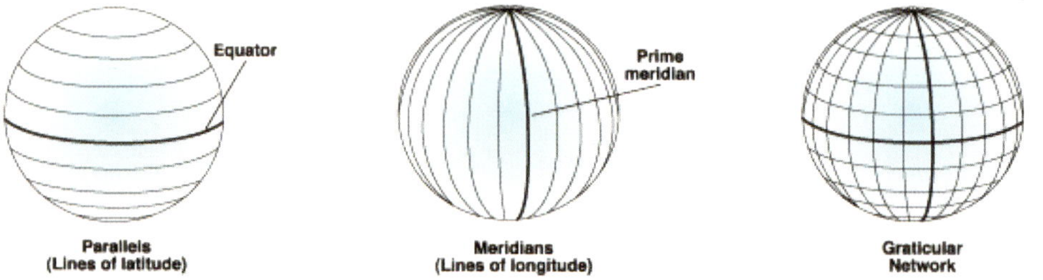

This figure shows the parallels and meridians that form a graticule.

Latitude and longitude values are traditionally measured either in decimal degrees or in degrees, minutes, and seconds (DMS). Latitude values are measured relative to the equator and range from -90° at the South Pole to +90° at the North Pole. Longitude values are measured relative to the prime meridian. They range from -180° when traveling west to 180° when traveling east. If the prime meridian is at Greenwich, then Australia, which is south of the equator and east of Greenwich, has positive longitude values and negative latitude values.

It may be helpful to equate longitude values with X and latitude values with Y. Data defined on a geographic coordinate system is displayed as if a degree is a linear unit of measure. This method is basically the same as the Plate Carrée projection.

Although longitude and latitude can locate exact positions on the surface of the globe, they are not uniform units of measure. Only along the equator does the distance represented by one degree of longitude approximate the distance represented by one degree of latitude. This is because the equator is the only parallel as large as a meridian. (Circles with the same radius as the spherical earth are called great circles. The equator and all meridians are great circles.)

Above and below the equator, the circles defining the parallels of latitude get gradually smaller until they become a single point at the North and South Poles where the meridians converge. As the meridians converge toward the poles, the distance represented by one degree of longitude decreases to zero. On the Clarke 1866 spheroid, one degree of longitude at the equator equals 111.321 km, while at 60° latitude it is only 55.802 km. Because degrees of latitude and longitude don't have a standard length, you can't measure distances or areas accurately or display the data easily on a flat map or computer screen.

Geographic Latitude and Longitude

Equator

The "latitude" (abbreviation: Lat., φ, or phi) of a point on Earth's surface is the angle between the equatorial plane and the straight line that passes through that point and through (or close to) the center of the Earth. Lines joining points of the same latitude trace circles on the surface of Earth called parallels, as they are parallel to the equator and to each other. The north pole is 90° N; the south pole is 90° S. The 0° parallel of latitude is designated the equator, the fundamental plane of all geographic coordinate systems. The equator divides the globe into Northern and Southern Hemispheres.

The "longitude" (abbreviation: Long., λ, or lambda) of a point on Earth's surface is the angle east or west of a reference meridian to another meridian that passes through that point. All meridians are halves of great ellipses (often called great circles), which converge at the north and south poles. The meridian of the British Royal Observatory in Greenwich, in south-east London, England, is the international prime meridian, although some organizations—such as the French Institut Géographique National—continue to use other meridians for internal purposes. The prime meridian determines the proper Eastern and Western Hemispheres, although maps often divide these hemispheres further west in order to keep the Old World on a single side. The antipodal meridian of Greenwich is both 180°W and 180°E. This is not to be conflated with the International Date Line, which diverges from it in several places for political reasons, including between far eastern Russia and the far western Aleutian Islands.

The combination of these two components specifies the position of any location on the surface of Earth, without consideration of altitude or depth. The grid formed by lines of latitude and longitude is known as a "graticule". The origin/zero point of this system is located in the Gulf of Guinea about 625 km (390 mi) south of Tema, Ghana.

Measuring Height using Datums

Complexity of the Problem

To completely specify a location of a topographical feature on, in, or above Earth, one also has to specify the vertical distance from Earth's center or surface.

Earth is not a sphere, but an irregular shape approximating a biaxial ellipsoid. It is nearly spherical, but has an equatorial bulge making the radius at the equator about 0.3% larger than the radius measured through the poles. The shorter axis approximately coincides with the axis of rotation. Though early navigators thought of the sea as a flat surface that could be used as a vertical datum, this is not actually the case. Earth has a series of layers of equal potential energy within its gravitational field. Height is a measurement at right angles to this surface, roughly toward Earth's centre, but local variations make the equipotential layers irregular (though roughly ellipsoidal). The choice of which layer to use for defining height is arbitrary.

Common Baselines

Common height baselines include

- The surface of the datum ellipsoid, resulting in an *ellipsoidal height*
- The mean sea level as described by the gravity geoid, yielding the orthometric height
- A vertical datum, yielding a dynamic height relative to a known reference height.

Along with the latitude ϕ and longitude λ, the height h provides the three-dimensional geodetic coordinates or geographic coordinates for a location.

Datums

In order to be unambiguous about the direction of "vertical" and the "surface" above which they are measuring, map-makers choose a reference ellipsoid with a given origin and orientation that best fits their need for the area they are mapping. They then choose the most appropriate mapping of the spherical coordinate system onto that ellipsoid, called a terrestrial reference system or geodetic datum.

Datums may be global, meaning that they represent the whole earth, or they may be local, meaning that they represent an ellipsoid best-fit to only a portion of the earth. Points on the earth's surface move relative to each other due to continental plate motion, subsidence, and diurnal movement caused by the moon and the tides. This daily movement can be as much as a metre. Continental movement can be up to 10 cm a year, or 10 m in a century. A weather system high-pressure area can cause a sinking of 5 mm. Scandinavia is rising by 1 cm a year as a result of the melting of the ice sheets of the last ice age, but neighbouring Scotland is rising by only 0.2 cm. These changes are insignificant if a local datum is used, but are statistically significant if a global datum is used.

Examples of global datums include World Geodetic System (WGS 84), the default datum used for the Global Positioning System, and the International Terrestrial Reference Frame (ITRF), used for estimating continental drift and crustal deformation. The distance to Earth's centre can be used both for very deep positions and for positions in space.

Local datums chosen by a national cartographical organisation include the North American Datum, the European ED50, and the British OSGB36. Given a location, the datum provides the latitude ϕ and longitude λ. In the United Kingdom there are three common latitude, longitude, and height systems in use. WGS 84 differs at Greenwich from the one used on published maps OSGB36 by approximately 112m. The military system ED50, used by NATO, differs from about 120m to 180m.

The latitude and longitude on a map made against a local datum may not be the same as one obtained from a GPS receiver. Coordinates from the mapping system can sometimes be roughly changed into another datum using a simple translation. For example, to convert from ETRF89 (GPS) to the Irish Grid add 49 metres to the east, and subtract 23.4 metres from the north. More generally one datum is changed into any other datum using a process called Helmert transformations. This involves converting the spherical coordinates into Cartesian coordinates and applying a seven parameter transformation (translation, three-dimensional rotation), and converting back.

In popular GIS software, data projected in latitude/longitude is often represented as a 'Geographic Coordinate System'. For example, data in latitude/longitude if the datum is the North American Datum of 1983 is denoted by 'GCS North American 1983'.

Map Projection

To establish the position of a geographic location on a map, a map projection is used to convert geodetic coordinates to two-dimensional coordinates on a map; it projects the datum ellipsoidal coordinates and height onto a flat surface of a map. The datum, along with a map projection applied to a grid of reference locations, establishes a *grid system* for plotting locations. Common map projections in current use include the Universal Transverse Mercator (UTM), the Military

Grid Reference System (MGRS), the United States National Grid (USNG), the Global Area Reference System (GARS) and the World Geographic Reference System (GEOREF). Coordinates on a map are usually in terms northing N and easting E offsets relative to a specified origin.

Map projection formulas depend in the geometry of the projection as well as parameters dependent on the particular location at which the map is projected. The set of parameters can vary based on type of project and the conventions chosen for the projection. For the transverse Mercator projection used in UTM, the parameters associated are the latitude and longitude of the natural origin, the false northing and false easting, and an overall scale factor. Given the parameters associated with particular location or grin, the projection formulas for the transverse Mercator are a complex mix of algebraic and trigonometric functions.

UTM and UPS Systems

The Universal Transverse Mercator (UTM) and Universal Polar Stereographic (UPS) coordinate systems both use a metric-based cartesian grid laid out on a conformally projected surface to locate positions on the surface of the Earth. The UTM system is not a single map projection but a series of sixty, each covering 6-degree bands of longitude. The UPS system is used for the polar regions, which are not covered by the UTM system.

Stereographic Coordinate System

During medieval times, the stereographic coordinate system was used for navigation purposes. The stereographic coordinate system was superseded by the latitude-longitude system. Although no longer used in navigation, the stereographic coordinate system is still used in modern times to describe crystallographic orientations in the fields of crystallography, mineralogy and materials science.

Cartesian Coordinates

Every point that is expressed in ellipsoidal coordinates can be expressed as an rectilinear x y z (Cartesian) coordinate. Cartesian coordinates simplify many mathematical calculations. The Cartesian systems of different datums are not equivalent.

Earth-centered, Earth-fixed

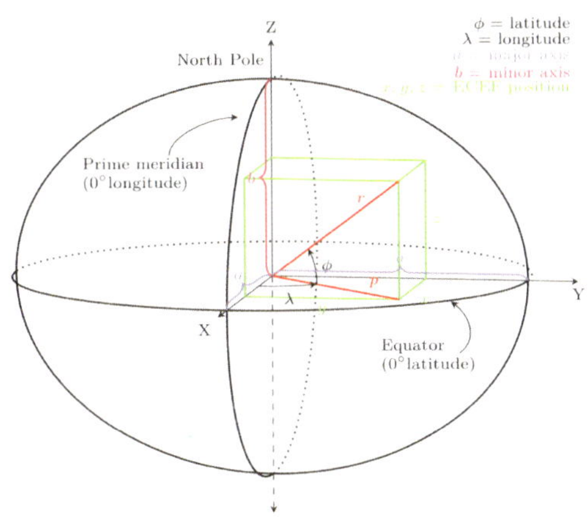

Earth Centered, Earth Fixed coordinates in relation to latitude and longitude.

The earth-centered earth-fixed (also known as the ECEF, ECF, or conventional terrestrial coordinate system) rotates with the Earth and has its origin at the center of the Earth.

The conventional right-handed coordinate system puts:

- The origin at the center of mass of the earth, a point close to the Earth's center of figure
- The Z axis on the line between the north and south poles, with positive values increasing northward (but does not exactly coincide with the Earth's rotational axis)
- The X and Y axes in the plane of the equator
- The X axis passing through extending from 180 degrees longitude at the equator (negative) to 0 degrees longitude (prime meridian) at the equator (positive)
- The Y axis passing through extending from 90 degrees west longitude at the equator (negative) to 90 degrees east longitude at the equator (positive)

An example is the NGS data for a brass disk near Donner Summit, in California. Given the dimensions of the ellipsoid, the conversion from lat/lon/height-above-ellipsoid coordinates to X-Y-Z is straightforward—calculate the X-Y-Z for the given lat-lon on the surface of the ellipsoid and add the X-Y-Z vector that is perpendicular to the ellipsoid there and has length equal to the point's height above the ellipsoid. The reverse conversion is harder: given X-Y-Z we can immediately get longitude, but no closed formula for latitude and height exists. Using Bowring's formula in 1976 *Survey Review* the first iteration gives latitude correct within 10^{-11} degree as long as the point is within 10000 meters above or 5000 meters below the ellipsoid.

Local East, North, Up (ENU) Coordinates

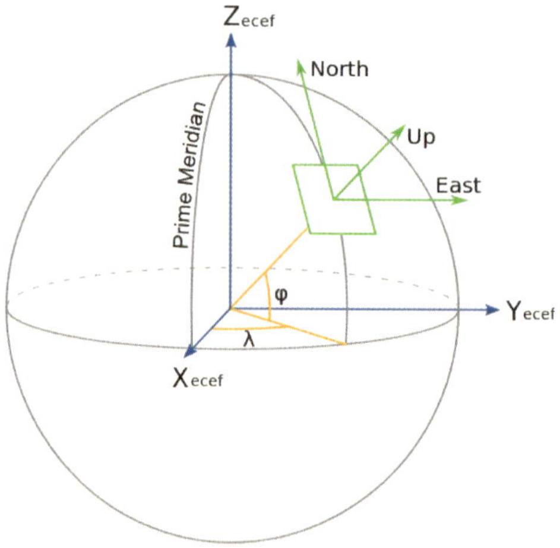

Earth Centred Earth Fixed and East, North, Up coordinates.

In many targeting and tracking applications the local East, North, Up (ENU) Cartesian coordinate system is far more intuitive and practical than ECEF or Geodetic coordinates. The local ENU coordinates are formed from a plane tangent to the Earth's surface fixed to a specific location and hence it is sometimes known as a "Local Tangent" or "local geodetic" plane. By convention the east axis is labeled x, the north y and the up z.

Local North, East, Down (NED) Coordinates

Also known as local tangent plane (LTP). In an airplane, most objects of interest are below the aircraft, so it is sensible to define down as a positive number. The North, East, Down (NED) coordinates allow this as an alternative to the ENU local tangent plane. By convention, the north axis is labeled x', the east y' and the down z'. To avoid confusion between x and x', etc. in this article we will restrict the local coordinate frame to ENU.

Expressing latitude and longitude as linear units

On the GRS80 or WGS84 spheroid at sea level at the equator, one latitudinal second measures *30.715 metres,* one latitudinal minute is *1843 metres* and one latitudinal degree is *110.6 kilometres*. The circles of longitude, meridians, meet at the geographical poles, with the west-east width of a second naturally decreasing as latitude increases. On the equator at sea level, one longitudinal second measures *30.92 metres,* a longitudinal minute is *1855 metres* and a longitudinal degree is *111.3 kilometres*. At 30° a longitudinal second is *26.76 metres,* at Greenwich (51°28′38″N) *19.22 metres,* and at 60° it is *15.42 metres*.

On the WGS84 spheroid, the length in meters of a degree of latitude at latitude φ (that is, the distance along a north–south line from latitude $(\varphi - 0.5)$ degrees to $(\varphi + 0.5)$ degrees) is about

$$111132.92 - 559.82\cos 2\varphi + 1.175\cos 4\varphi - 0.0023\cos 6\varphi$$

Similarly, the length in meters of a degree of longitude can be calculated as

$$111412.84\cos\varphi - 93.5\cos 3\varphi + 0.118\cos 5\varphi$$

(Those coefficients can be improved, but as they stand the distance they give is correct within a centimeter.)

An alternative method to estimate the length of a longitudinal degree at latitude is to assume a spherical Earth (to get the width per minute and second, divide by 60 and 3600, respectively):

$$\frac{\pi}{180} M_r \cos\varphi$$

where Earth's average meridional radius M_r is 6,367,449 m. Since the Earth is not spherical that result can be off by several tenths of a percent; a better approximation of a longitudinal degree at latitude φ is

$$\frac{\pi}{180} a\cos\beta$$

where Earth's equatorial radius a equals 6,378,137 m and $\tan\beta = \frac{b}{a}\tan\varphi$; for the GRS80 and WGS84 spheroids, b/a calculates to be 0.99664719. (β is known as the reduced (or parametric) latitude). Aside from rounding, this is the exact distance along a parallel of latitude; getting the distance along the shortest route will be more work, but those two distances are always within 0.6 meter of each other if the two points are one degree of longitude apart.

Longitudinal length equivalents at selected latitudes					
Latitude	City	Degree	Minute	Second	±0.0001°
60°	Saint Petersburg	55.80 km	0.930 km	15.50 m	5.58 m
51° 28′ 38″ N	Greenwich	69.47 km	1.158 km	19.30 m	6.95 m
45°	Bordeaux	78.85 km	1.31 km	21.90 m	7.89 m
30°	New Orleans	96.49 km	1.61 km	26.80 m	9.65 m
0°	Quito	111.3 km	1.855 km	30.92 m	11.13 m

Geostationary Coordinates

Geostationary satellites (e.g., television satellites) are over the equator at a specific point on Earth, so their position related to Earth is expressed in longitude degrees only. Their latitude is always zero (or approximately so), that is, over the equator.

Geodetic Datum

A geodetic datum is a tool used to define the shape and size of the earth, as well as the reference point for the various coordinate systems used in mapping the earth. Throughout time, hundreds of different datums have been used - each one changing with the earth views of the times.

True geodetic datums, however, are only those which appeared after the 1700s. Prior to that, the earth's ellipsoidal shape was not always taken into consideration, as many still believed it was flat. Since most datums today are used for measuring and showing large portions of the earth, an ellipsoidal model is essential.

The Vertical and Horizontal Datums

Today, there are hundreds of different datums in use; but, they are all either horizontal or vertical in their orientation.

The horizontal datum is the one that is used in measuring a specific position on the earth's surface in coordinate systems such as latitude and longitude. Because of the different local datums (i.e. those having different reference points), the same position can have many different geographic coordinates so it is important to know which datum the reference is in.

The vertical datum measures the elevations of specific points on the earth. This data is gathered via tides with sea level measurements, geodetic surveying with different ellipsoid models used with the horizontal datum, and gravity, measured with the geoid. The data is then depicted on maps as some height above sea level.

For reference, the geoid is a mathematical model of the earth measured with gravity that corresponds with the mean ocean surface level on the earth- such as if the water were extended over the land. Because the surface is highly irregular, however, there are different local geoids that are used to get the most accurate mathematical model possible for use in measuring vertical distances.

Commonly Used Datums

As previously mentioned, there are many datums in use around the world today. Some of the most commonly used datums are those of the World Geodetic System, the North American Datums, those of the Ordinance Survey of Great Britain, and the European Datum; however, this is by no means an exhaustive list.

Within the World Geodetic System (WGS), there are several different datums that have been in use throughout the years. These are WGS 84, 72, 70, and 60. The WGS 84 is currently the one in use for this system and is valid until 2010. In addition, it is one of the most widely used datums around the world.

In the 1980s, the United States Department of Defense used the Geodetic Reference System, 1980 (GRS 80) and Doppler satellite images to create a new, more accurate world geodetic system. This became what is known today as WGS 84. In terms of reference, WGS 84 uses what is called the "zero meridian" but because of the new measurements, it shifted 100 meters (0.062 miles) from the previously used Prime Meridian.

Similar to WGS 84 is the North American Datum 1983 (NAD 83). This is the official horizontal datum for use in the North and Central American geodetic networks. Like WGS 84, it is based on the GRS 80 ellipsoid so the two have very similar measurements. NAD 83 was also developed using satellite and remote sensing imagery and is the default datum on most GPS units today.

Prior to NAD 83 was NAD 27, a horizontal datum constructed in 1927 based on the Clarke 1866 ellipsoid. Though NAD 27 was in use for many years and still appears on the United States topographic maps, it was based on a series of approximations with the geodetic center being based at Meades Ranch, Kansas. This point was chosen because it is near the geographic center of the contiguous United States.

Also similar to WGS 84 is the Ordinance Survey of Great Britain 1936 (OSGB36) as the latitude and longitude positions of points are the same in both datums. However, it is based on the Airy 1830 ellipsoid as it shows Great Britain, its primary user, the most accurately.

The European Datum 1950 (ED50) is the datum used for showing much of Western Europe and was developed after World War II when a reliable system of mapping borders was needed. It was based on the International Ellipsoid but changed when GRS80 and WGS84 were put into use. Today ED50's latitude and longitude lines are similar to WGS84 but the lines do become farther apart on ED50 when moving toward Eastern Europe.

When working with these or other map datums, it is important to always be aware of which datum a particular map is referenced in because often there are large differences in terms of distance between place to place on each different datum. This "datum shift" can then cause problems in terms of navigation and/or in trying to locate a specific place or object as a user of the wrong datum can sometimes be hundreds of meters from their desired position.

Whichever datum is used, however, they represent a powerful geographic tool but are most important in cartography, geology, navigation, surveying, and sometimes even astronomy. In fact, "geodesy" (the study of measurement and Earth representation) has become its own subject within the field of earth sciences.

Geodetic Coordinates

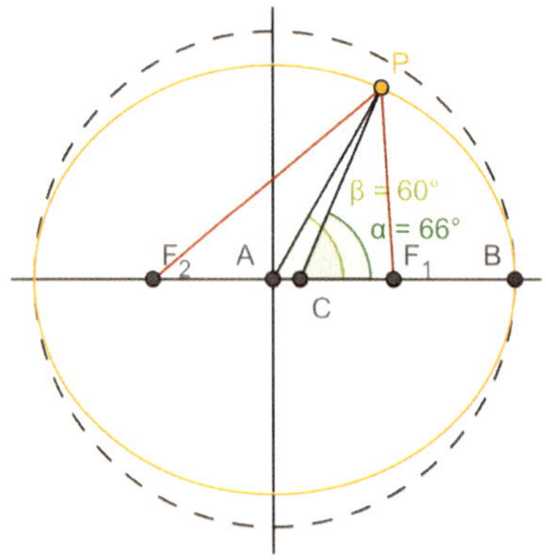

The same position on a spheroid has a different angle for latitude depending on whether the angle is measured from the normal line segment *IP* of the ellipsoid (angle α) or the line segment *AP* from the center (angle β). Note that the "flatness" of the spheroid (orange) in the image is greater than that of the Earth; as a result, the corresponding difference between the "geodetic" and "geocentric" latitudes is also exaggerated.

In geodetic coordinates, the Earth's surface is approximated by an ellipsoid, and locations near the surface are described in terms of latitude (ϕ), longitude (λ) and height (h).

Geodetic Versus Geocentric Latitude

It is important to note that geodetic latitude (ϕ), resp. altitude, is different from geocentric latitude (ϕ), resp. altitude. Geodetic latitude is determined by the angle between the equatorial plane and normal to the ellipsoid, whereas geocentric latitude is determined by the angle between the equatorial plane and line joining the point to the centre of the ellipsoid. Unless otherwise specified latitude is geodetic latitude.

Earth Reference Ellipsoid

Defining and Derived Parameters

The ellipsoid is completely parameterized by the semi-major axis a and the flattening f.

Parameter	Symbol
Semi-major axis	a
Reciprocal of flattening	$\frac{1}{f}$

From a and f it is possible to derive the semi-minor axis b, first eccentricity e and second eccentricity $e'$of the ellipsoid

Parameter	Value
Semi-minor axis	$b = a(1 - f)$

First eccentricity squared	$e^2 = 1 - \frac{b^2}{a^2} = 2f(1-f)$
Second eccentricity squared	$e'^2 = \frac{a^2}{b^2} - 1 = \frac{f(2-f)}{(1-f)^2}$

Parameters for Some Geodetic Systems

Australian Geodetic Datum 1966 [AGD66] and Australian Geodetic Datum 1984 (AGD84)

AGD66 and AGD84 both use the parameters defined by Australian National Spheroid.

Australian National Spheroid (ANS)

ANS defining parameters		
Parameter	Notation	Value
Semi-major axis	a	6378160.000 m
Reciprocal of flattening	$\frac{1}{f}$	298.25

Geocentric Datum of Australia 1994 (GDA94)

GDA94 uses the parameters defined by GRS80

Geodetic Reference System 1980 (GRS80)

GRS80 parameters		
Parameter	Notation	Value
Semi-major axis	a	6378137 m
Reciprocal of flattening	$\frac{1}{f}$	298.257222101

See GDA Technical Manual document for more details; the value given above for the flattening is not exact.

World Geodetic System 1984 (WGS 84)

The Global Positioning System (GPS) uses the World Geodetic System 1984 (WGS 84) to determine the location of a point near the surface of the Earth.

WGS 84 defining parameters		
Parameter	Notation	Value
Semi-major axis	a	6378137.0 m

Reciprocal of flattening	$\frac{1}{f}$	298.257223563

WGS 84 derived geometric constants		
Constant	Notation	Value
Semi-minor axis	b	6356752.3142 m
First eccentricity squared	e^2	$6.69437999014\times10^{-3}$
Second eccentricity squared	e'^2	$6.73949674228\times10^{-3}$

Conversion Calculations

Datum conversion is the process of converting the coordinates of a point from one datum system to another. Datum conversion may frequently be accompanied by a change of grid projection.

Examples

Examples of map datums are:

- WGS 84, 72, 64 and 60 of the World Geodetic System
- NAD83, the North American Datum which is very similar to WGS 84
- NAD27, the older North American Datum, of which NAD83 was basically a readjustment
- OSGB36 of the Ordnance Survey of Great Britain
- ED50, the European Datum
- GDA94, the Australian Datum
- JGD2011, the Japanese Datum, adjusted for changes caused by 2011 Tōhoku earthquake and tsunami
- Tokyo97, the older Japanese Datum
- KGD2002, the Korean Datum
- TWD67 and TWD97, different datum currently used in Taiwan.
- BJS54 and XAS80, old geodetic datum used in China
- GCJ-02 and BD-09, Chinese encrypted geodetic datum.
- PZ-90.11, the current geodetic reference used by GLONASS
- GTRF, the geodetic reference used by Galileo
- CGCS2000, or CGS-2000, the geodetic reference used by BeiDou Navigation Satellite System

- International Terrestrial Reference Frames (ITRF88, 89, 90, 91, 92, 93, 94, 96, 97, 2000, 2005, 2008, 2014), different realizations of the ITRS.
- Hong Kong Principal Datum, a vertical datum used in Hong Kong.

Figure of the Earth

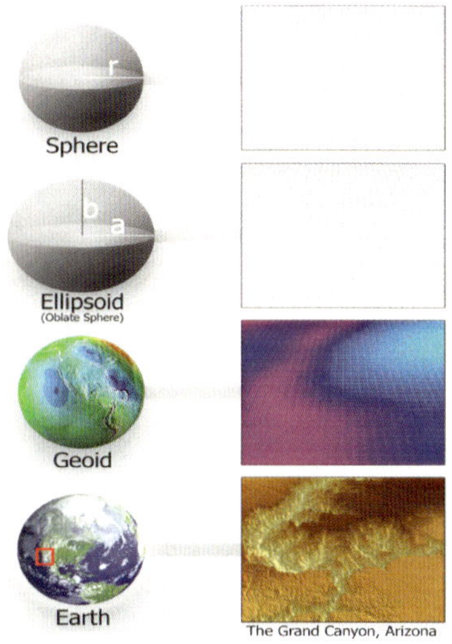

Because the surface of the Earth is so complex, geodesists use simplified, mathematical models of the Earth for many applications.

The Earth's shape is nearly spherical, with a radius of about 3,963 miles (6,378 km), and its surface is very irregular. Mountains and valleys make actually measuring this surface impossible because an infinite amount of data would be needed. For example, if you wanted to find the actual surface area of the Grand Canyon, you would have to cover every inch of land. It would take you many lifetimes to measure every crevice, valley, and rise. You could never complete the project because it would take too long.

To measure the Earth and avoid the problems that places like the Grand Canyon present, geodesists use a theoretical mathematical surface called the ellipsoid. Because the ellipsoid exists only in theory and not in real life, it can be completely smooth and does not take any irregularities - such as mountains or valleys -- into account. The ellipsoid is created by rotating an ellipse around its shorter axis. This matches the real Earth's shape, because the earth is slightly flattened at the poles and bulges at the equator.

While the ellipsoid gives a common reference to geodesists, it is still only a mathematical concept. Geodesists often need to account for the reality of the Earth's surface. To meet this need, the geoid, a shape that refers to global mean sea level, was created. If the geoid really existed, the surface of the Earth would be equal to a level in between the high-tide and low-tide marks.

Although a geoid may seem to be a smooth, regular shape, it isn't. The Earth's mass is unevenly distributed, meaning that certain areas of the planet experience more gravitational "pull" than others. Because of these variations in gravitational force, the "height" of different parts of the geoid is always changing, moving up and down in response to gravity. The geoidal surface is an irregular shape with a wavy appearance; there are rises in some areas and dips in others.

Need for Models of the Figure of the Earth

The actual topographic surface is most apparent with its variety of land forms and water areas. This is, in fact, the surface on which actual Earth measurements are made. However, it is not feasible for exact mathematical analysis, because the formulas which would be required to take the irregularities into account would necessitate a prohibitive amount of computation. The topographic surface is generally the concern of topographers and hydrographers.

The Pythagorean concept of a spherical Earth offers a simple surface that is mathematically easy to deal with. Many astronomical and navigational computations use it as a surface representing the Earth. While the sphere is a close approximation of the true figure of the Earth and satisfactory for many purposes, to the geodesists interested in the measurement of long distances on the scale of continents and oceans, a more exact figure is necessary. Closer approximations range from modelling the shape of the surface of the entire Earth as an oblate spheroid or an oblate ellipsoid, to the use of spherical harmonics to approximate the geoid, or local approximations in terms of local reference ellipsoids.

The idea of a planar or flat surface for Earth, however, is still sufficient for surveys of small areas, as the local topography is far more significant than the curvature. Plane-table surveys are made for relatively small areas, and no account is taken of the curvature of the Earth. A survey of a city would likely be computed as though the Earth were a plane surface the size of the city. For such small areas, nearly exact positions can be determined relative to each other without considering the size and shape of the entire Earth.

In the mid- to late 20th century, research across the geosciences contributed to drastic improvements in the accuracy of the figure of the Earth. The primary utility (and the motivation for funding, mainly from the military) of this improved accuracy was to provide geographical and gravitational data for the inertial guidance systems of ballistic missiles. This funding also drove the expansion of geoscientific disciplines, fostering the creation and growth of various geoscience departments at many universities.

Models of the Figure of the Earth

The models for the figure of the Earth vary in the way they are used, in their complexity, and in the accuracy with which they represent the size and shape of the Earth.

Sphere

The simplest model for the shape of the entire Earth is a sphere. The Earth's radius is the distance from Earth's center to its surface, about 6,371 kilometers (3,959 mi). While "radius" normally is a characteristic of perfect spheres, the Earth deviates from spherical by only a third of a percent, sufficiently close to treat it as a sphere in many contexts and justifying the term "the radius of the Earth".

The concept of a spherical Earth dates back to around the 6th century BC, but remained a matter of philosophical speculation until the 3rd century BC. The first scientific estimation of the radius of the earth was given by Eratosthenes about 240 BC, with estimates of the accuracy of Eratosthenes's measurement ranging from 2% to 15%.

A view across a 20-km-wide bay in the coast of Spain. Note the curvature of the Earth hiding the base of the buildings on the far shore.

The Earth is only approximately spherical, so no single value serves as its natural radius. Distances from points on the surface to the center range from 6,353 km to 6,384 km (3,947 – 3,968 mi). Several different ways of modeling the Earth as a sphere each yield a mean radius of 6,371 kilometers (3,959 mi). Regardless of the model, any radius falls between the polar minimum of about 6,357 km and the equatorial maximum of about 6,378 km (3,950 – 3,963 mi). The difference 21 kilometers (13 mi) correspond to the polar radius being approximately 0.3% shorter than the equator radius.

Ellipsoid of Revolution

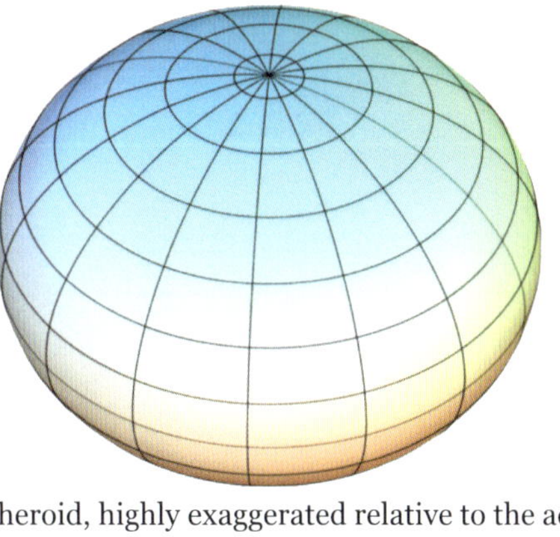

An oblate spheroid, highly exaggerated relative to the actual Earth

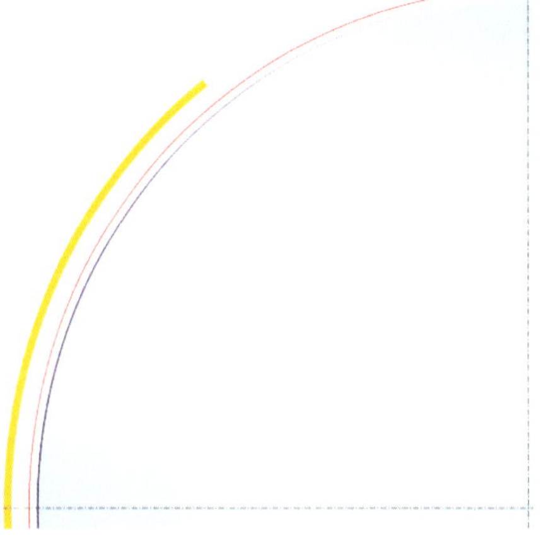

A scale diagram of the oblateness of the 2003 IERS reference ellipsoid, with north at the top. The outer edge of the dark blue line is an ellipse with the same eccentricity as that of Earth. For comparison, the light blue circle within has a diameter equal to the ellipse's minor axis. The red curve represents the Karman line 100 km (62 mi) above sea level, while the yellow band denotes the altitude range of the ISS in low Earth orbit.

Since the Earth is flattened at the poles and bulges at the Equator, geodesy represents the figure of the Earth as an oblate spheroid. The oblate spheroid, or oblate ellipsoid, is an ellipsoid of revolution obtained by rotating an ellipse about its shorter axis. It is the regular geometric shape that most nearly approximates the shape of the Earth. A spheroid describing the figure of the Earth or other celestial body is called a reference ellipsoid. The reference ellipsoid for Earth is called an Earth ellipsoid.

An ellipsoid of revolution is uniquely defined by two quantities. Several conventions for expressing the two quantities are used in geodesy, but they are all equivalent to and convertible with each other:

- Equatorial radius a (called semi-major axis), and polar radius b (called semi-minor axis);
- a and eccentricity e;
- a and flattening f.

Eccentricity and flattening are different ways of expressing how squashed the ellipsoid is. When flattening appears as one of the defining quantities in geodesy, generally it is expressed by its reciprocal. For example, in the WGS 84 spheroid used by today's GPS systems, the reciprocal of the flattening $\frac{1}{f}$ is set to be exactly 298.257223563.

The difference between a sphere and a reference ellipsoid for Earth is small, only about one part in 300. Historically, flattening was computed from grade measurements. Nowadays, geodetic networks and satellite geodesy are used. In practice, many reference ellipsoids have been developed over the centuries from different surveys. The flattening value varies slightly from one reference ellipsoid to another, reflecting local conditions and whether the reference ellipsoid is intended to model the entire Earth or only some portion of it.

A sphere has a single radius of curvature, which is simply the radius of the sphere. More complex surfaces have radii of curvature that vary over the surface. The radius of curvature describes the radius of the sphere that best approximates the surface at that point. Oblate ellipsoids have constant radius of curvature east to west along parallels, if a graticule is drawn on the surface, but varying curvature in any other direction. For an oblate ellipsoid, the polar radius of curvature r_p is larger than the equatorial

$$r_p = \frac{a^2}{b},$$

because the pole is flattened: the flatter the surface, the larger the sphere must be to approximate it. Conversely, the ellipsoid's north-south radius of curvature at the equator r_e is smaller than the polar

$$r_e = \frac{b^2}{a}$$

where a is the distance from the center of the ellipsoid to the equator (semi-major axis), and b is the distance from the center to the pole. (semi-minor axis).

More Complicated Shapes

The possibility that the Earth's equator is an ellipse rather than a circle and therefore that the ellipsoid is triaxial has been a matter of scientific controversy for many years. Modern technological developments have furnished new and rapid methods for data collection and, since the launch of *Sputnik 1,* orbital data have been used to investigate the theory of ellipticity.

A second theory, more complicated than triaxiality, proposed that observed long periodic orbital variations of the first Earth satellites indicate an additional depression at the south pole accompanied by a bulge of the same degree at the north pole. It is also contended that the northern middle latitudes were slightly flattened and the southern middle latitudes bulged in a similar amount. This concept suggested a slightly pear-shaped Earth and was the subject of much public discussion. Modern geodesy tends to retain the ellipsoid of revolution as a reference ellipsoid and treat triaxiality and pear shape as a part of the geoid figure: they are represented by the spherical harmonic coefficients C_{22}, S_{22} and C_{30}, respectively, corresponding to degree and order numbers 2.2 for the triaxiality and 3.0 for the pear shape.

Geoid

It was stated earlier that measurements are made on the apparent or topographic surface of the Earth and it has just been explained that computations are performed on an ellipsoid. One other surface is involved in geodetic measurement: the geoid. In geodetic surveying, the computation of the geodetic coordinates of points is commonly performed on a reference ellipsoid closely approximating the size and shape of the Earth in the area of the survey. The actual measurements made on the surface of the Earth with certain instruments are however referred to the geoid. The ellipsoid is a mathematically defined regular surface with specific dimensions. The geoid, on the other hand, coincides with that surface to which the oceans would conform over the entire Earth if free to adjust to the combined effect of the Earth's mass attraction (gravitation) and the centrifugal force of the Earth's rotation. As a result of the uneven distribution of the Earth's mass, the geoidal surface is irregular and, since the ellipsoid is a regular surface, the separations between the two, referred to as geoid undulations, geoid heights, or geoid separations, will be irregular as well.

The geoid is a surface along which the gravity potential is everywhere equal and to which the direction of gravity is always perpendicular. The latter is particularly important because optical instruments containing gravity-reference leveling devices are commonly used to make geodetic measurements. When properly adjusted, the vertical axis of the instrument coincides with the direction of gravity and is, therefore, perpendicular to the geoid. The angle between the plumb line which is perpendicular to the geoid (sometimes called "the vertical") and the perpendicular to the ellipsoid (sometimes called "the ellipsoidal normal") is defined as the deflection of the vertical. It has two components: an east-west and a north-south component.

Earth rotation and Earth's interior

Determining the exact figure of the Earth is not only a geodetic operation or a task of geometry, but is also related to geophysics. Without any idea of the Earth's interior, we can state a "constant density" of 5.515 g/cm³ and, according to theoretical arguments, such a body rotating like the Earth would have a flattening of 1:230.

In fact, the measured flattening is 1:298.25, which is closer to a sphere and a strong argument that the Earth's core is *very compact*. Therefore, the density must be a function of the depth, ranging from 2.6 g/cm³ at the surface (rock density of granite, etc.), up to 13 g/cm³ within the inner core, see Structure of the Earth.

Volume

Earth is an ellipsoid, so its volume is $V = \frac{4}{3\pi}$ abc, where a, b and c are its three orthogonal principal semi-axes, or radii. Using the parameters of the WGS84 ellipsoid of revolution, an oblate spheroid where $a = b > c, a = b = 6{,}378.137$ km and

$$c = 6{,}357.7523142 km, \;\; V = 1.0832073198 \times 10^{\wedge 12}\, km^3 \approx 1.08321 \times 10^{12}\, km^3 \left(2.5988 \times 10^{11}\, cu\, mi\right).$$

Gravity

Imagine if all the mountains and valleys were scoured off the planet leaving a continuous world ocean completely at rest. The effects of the Earth's gravity on this hypothetical world mean sea level is represented by the geoid.

Gravity is the force that pulls all objects in the universe toward each other. On Earth, gravity pulls all objects "downward" toward the center of the planet. According to Sir Isaac Newton's Universal Law of Gravitation, the gravitational attraction between two bodies is stronger when the masses of the objects are greater and closer together. This rule applies to the Earth's gravitational field as well. Because the Earth rotates and its mass and density vary at different locations on the planet, gravity also varies.

One reason that geodesists measure variations in the Earth's gravity is because gravity plays a major role in determining mean sea level. Geodesists calculate the elevation of locations on the Earth's surface based on the mean sea level. So knowing how gravity changes sea level helps geodesists make more accurate measurements. In general, in areas of the planet where gravitational forces are stronger, the mean sea level will be higher. In areas where the Earth's gravitational forces are weaker, the mean sea level will be lower.

To measure the Earth's gravity field, geodesists use instruments in space and on land. In space, satellites gather data on gravitational changes as they pass over points on the Earth's surface. On land, devices called gravimeters measure the Earth's gravitational pull on a suspended mass. With this data, geodesists can create detailed maps of gravitational fields and adjust elevations on existing maps. Gravity principally affects the vertical datum because it changes the elevation of the land surface.

Cartesian Coordinate System

Cartesian coordinates, also called rectangular coordinates, provide a method of rendering graphs and indicating the positions of points on a two-dimensional (2D) surface or in three-dimensional (3D) space. The scheme gets its name from one of the first people known to have used it, the French mathematician and philosopher René Descartes (1596-1650). Cartesian coordinates are used to define positions on computer displays, in 3D models and virtual reality (VR) renderings. The coordinate system is also employed in mathematics, physics, engineering, navigation, robotics, economics and other sciences.

The Cartesian plane consists of two perpendicular axes that cross at a central point called the origin. Positions or coordinates are determined according to the *east/west* and *north/south* displacements from the origin. The east/west axis is often called the x axis, and the north/south axis is called the y axis. For this reason, the Cartesian plane is also known as the xy -plane. The x and y axes are linear number lines, meaning that each division on a given axis always represents the same increment. However, the increments on different axes can differ. For example, in the illustration at left below, each increment on the x axis might represent 2 units, while each increment on the y axis represents 5 units. Points or coordinates are indicated by writing an opening parenthesis, the x value, a comma, the y value, and a closing parenthesis in that order. An example is (x,y) = (2,-5). The origin is usually, but not always, assigned the value (0,0).

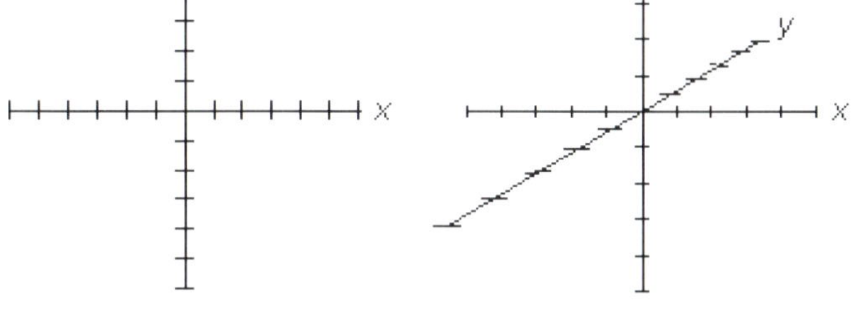

Cartesian three-space, also called xyz -space, has a third axis, oriented at right angles to the xy plane. This axis, usually called the z axis, passes through the origin of the xy -plane. Positions or coordinates are determined according to the east/west (x), north/south (y), and *up/down* (z) displacements from the origin. As is the case with the x and y axes, the z axis is a linear number line. For example, in the illustration at right above, each increment on the x axis might represent 5 units, each increment on the y axis 10 units, and each increment on the z axis 2 units. Points or coordinates are indicated by writing an opening parenthesis, the x value, a comma, the y value, another comma, the *z* value, and a closing parenthesis in that order. An example is (*x*,*y*,*z*) = (5,-10,-4). The origin is usually, but not always, assigned the value (0,0,0).

Using the Cartesian coordinate system, geometric shapes (such as curves) can be described by algebraic equations, namely equations satisfied by the coordinates of the points lying on the shape. For example, a circle of radius 2 may be described by the equation $x^2 + y^2 = 4$.

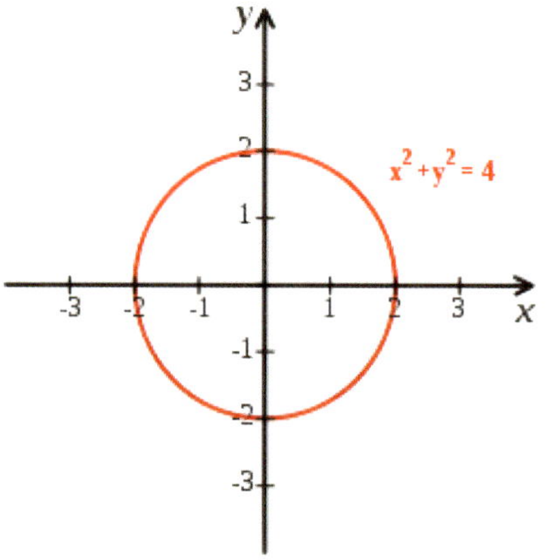

Cartesian coordinate system with the circle of radius 2 centered at the origin marked in red. The equation of the circle is $x^2 + y^2 = 4$.

Description

One Dimension

Choosing a Cartesian coordinate system for a one-dimensional space – that is, for a straight line—involves choosing a point *O* of the line (the origin), a unit of length, and an orientation for the line. An orientation chooses which of the two half-lines determined by *O* is the positive, and which is negative; we then say that the line "is oriented" (or "points") from the negative half towards the positive half. Then each point *P* of the line can be specified by its distance from *O*, taken with a + or – sign depending on which half-line contains *P*.

A line with a chosen Cartesian system is called a number line. Every real number has a unique location on the line. Conversely, every point on the line can be interpreted as a number in an ordered continuum such as the real numbers.

Two Dimensions

A Cartesian coordinate system in two dimensions (also called a rectangular coordinate system or an orthogonal coordinate system) is defined by an ordered pair of perpendicular lines (axes), a single unit of length for both axes, and an orientation for each axis. The point where the axes meet is taken as the origin for both, thus turning each axis into a number line. For any point *P*, a line is drawn through *P* perpendicular to each axis, and the position where it meets the axis is interpreted as a number. The two numbers, in that chosen order, are the *Cartesian coordinates* of *P*. The reverse construction allows one to determine the point *P* given its coordinates.

The first and second coordinates are called the *abscissa* and the *ordinate* of *P*, respectively; and the point where the axes meet is called the *origin* of the coordinate system. The coordinates are usually written as two numbers in parentheses, in that order, separated by a comma, as in (3, -10.5). Thus the origin has coordinates (0,0), and the points on the positive half-axes, one unit away from the origin, have coordinates (1,0) and (0,1).

In mathematics, physics, and engineering, the first axis is usually defined or depicted as horizontal and oriented to the right, and the second axis is vertical and oriented upwards. (However, in some computer graphics contexts, the ordinate axis may be oriented downwards.) The origin is often labeled *O*, and the two coordinates are often denoted by the letters *X* and *Y*, or *x* and *y*. The axes may then be referred to as the *X*-axis and *Y*-axis. The choices of letters come from the original convention, which is to use the latter part of the alphabet to indicate unknown values. The first part of the alphabet was used to designate known values.

A Euclidean plane with a chosen Cartesian coordinate system *Cartesian plane*. In a Cartesian plane one can define canonical representatives of certain geometric figures, such as the unit circle (with radius equal to the length unit, and center at the origin), the unit square (whose diagonal has endpoints at (0,0) and (1,1)), the unit hyperbola, and so on.

The two axes divide the plane into four right angles, called *quadrant*. The quadrants may be named or numbered in various ways, but the quadrant where all coordinates are positive is usually called the *first quadrant*.

If the coordinates of a point are (x,y), then its distances from the X-axis and from the Y-axis are $|y|$ and $|x|$, respectively; where $|....|$ denotes the absolute value of a number.

Three Dimensions

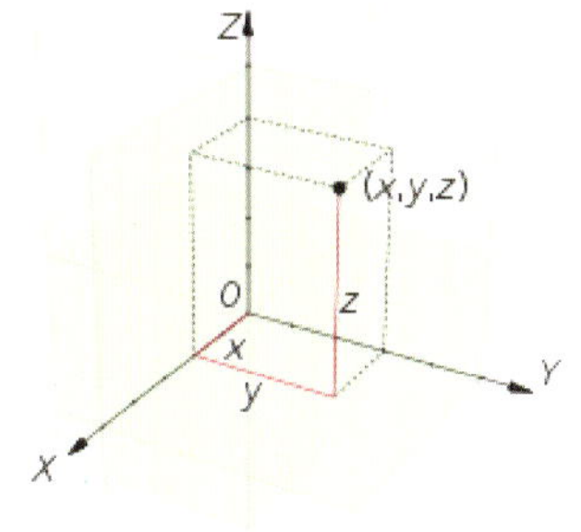

A three dimensional Cartesian coordinate system, with origin ***O*** and axis lines ***X***, ***Y*** and ***Z***, oriented as shown by the arrows. The tick marks on the axes are one length unit apart. The black dot shows the point with coordinates $x = 2$, $y = 3$, and $z = 4$, or (2, 3, 4).

A Cartesian coordinate system for a three-dimensional space consists of an ordered triplet of lines (the *axes*) that go through a common point (the *origin*), and are pair-wise perpendicular; an orientation for each axis; and a single unit of length for all three axes. As in the two-dimensional case, each axis becomes a number line. For any point *P* of space, one considers a plane through *P* perpendicular to each coordinate axis, and interprets the point where that plane cuts the axis as a number. The Cartesian coordinates of *P* are those three numbers, in the chosen order The reverse construction determines the point *P* given its three coordinates.

Alternatively, each coordinate of a point *P* can be taken as the distance from *P* to the plane defined by the other two axes, with the sign determined by the orientation of the corresponding axis.

Each pair of axes defines a *coordinate plane*. These planes divide space into eight trihedral, called *octants*.

The coordinates are usually written as three numbers (or algebraic formulas) surrounded by parentheses and separated by commas, as in (3,-2.5,1) or $(t,u+v,\pi/2)$. Thus, the origin has coordinates (0,0,0), and the unit points on the three axes are (1,0,0), (0,1,0), and (0,0,1).

There are no standard names for the coordinates in the three axes. The coordinates are often denoted by the letters *X*, *Y*, and *Z* (or *x*, *y*, and *z*), in which case the lines are called the *X*-, *Y*-, and *Z*-axis, respectively. Then the coordinate planes can be referred to as the *XY*-, *YZ*-, and *XZ*-planes.

In mathematics, physics, and engineering contexts, the first two axes are often defined or depicted as horizontal, with the third axis pointing up. In that case the third coordinate may be called *height* or *altitude*. The orientations are usually chosen so that the 90 degree angle from the first axis to the second axis looks counter-clockwise when seen from the point (0,0,1); a convention that is commonly called *the right hand rule*.

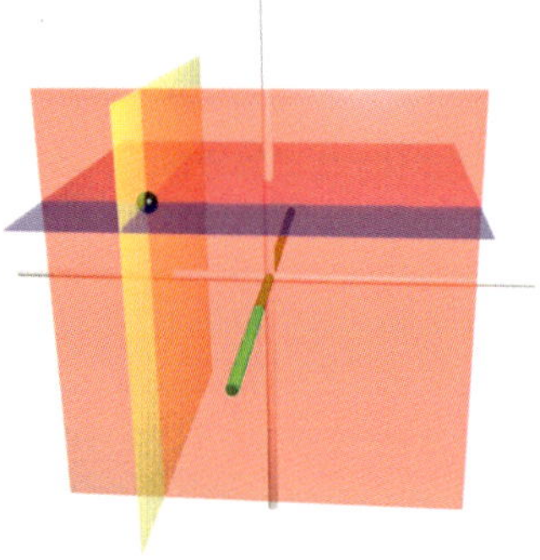

The coordinate surfaces of the Cartesian coordinates (x, y, z). The z-axis is vertical and the x-axis is highlighted in green. Thus, the red plane shows the points with $x = 1$, the blue plane shows the points with $z = 1$, and the yellow plane shows the points with $y = -1$. The three surfaces intersect at the point *P* (shown as a black sphere) with the Cartesian coordinates (1, −1, 1).

Higher Dimensions

A Euclidean plane with a chosen Cartesian system is called a Cartesian plane. Since Cartesian coordinates are unique and non-ambiguous, the points of a Cartesian plane can be identified with pairs of real numbers; that is with the Cartesian product $\mathbb{R}^2 = \mathbb{R} \times \mathbb{R}$, , where $\mathbb{R}$ is the set of all real number. In the same way, the points in any Euclidean space of dimension n be identified with the tuples (lists) of n real numbers, that is, with the Cartesian product $\mathbb{R}^n$.

Generalizations

The concept of Cartesian coordinates generalizes to allow axes that are not perpendicular to each other, and/or different units along each axis. In that case, each coordinate is obtained by projecting the point onto one axis along a direction that is parallel to the other axis (or, in general, to the hyper-plane defined by all the other axes). In such an oblique coordinate system the computations of distances and angles must be modified from that in standard Cartesian systems, and many standard formulas (such as the Pythagorean formula for the distance) do not hold.

Notations and Conventions

The Cartesian coordinates of a point are usually written in parentheses and separated by commas, as in (10, 5) or (3, 5, 7). The origin is often labelled with the capital letter *O*. In analytic geometry, unknown or generic coordinates are often denoted by the letters (x, y) in the plane, and (x, y, z) in three-dimensional space. This custom comes from a convention of algebra, which uses letters near the end of the alphabet for unknown values (such as were the coordinates of points in many geometric problems), and letters near the beginning for given quantities.

These conventional names are often used in other domains, such as physics and engineering, although other letters may be used. For example, in a graph showing how a pressure varies with time, the graph coordinates may be denoted *p* and *t*. Each axis is usually named after the coordinate which is measured along it; so one says the *x-axis*, the *y-axis*, the *t-axis*, etc.

Another common convention for coordinate naming is to use subscripts, as $(x_1, x_2, \ldots, x_n)$ for the n coordinates in an n-dimensional space, especially when n is greater than 3 or unspecified. Some authors prefer the numbering $(x_0, x_1, \ldots, x_{n-1})$. These notations are especially advantageous in computer programming: by storing the coordinates of a point as an array, instead of a record, the subscript can serve to index the coordinates.

In mathematical illustrations of two-dimensional Cartesian systems, the first coordinate (traditionally called the abscissa) is measured along a horizontal axis, oriented from left to right. The second coordinate (the ordinate) is then measured along a vertical axis, usually oriented from bottom to top. Young children learning the Cartesian system, commonly learn the order to read the values before cementing the *x*-, *y*-, and *z*-axis concepts, by starting with 2D mnemonics (e.g. 'Walk along the hall then up the stairs' akin to straight across the *x*-axis then up vertically along the *y*-axis).

Computer graphics and image processing, however, often use a coordinate system with the *y*-axis oriented downwards on the computer display. This convention developed in the 1960s (or earlier) from the way that images were originally stored in display buffers.

For three-dimensional systems, a convention is to portray the *xy*-plane horizontally, with the *z*-ax- is added to represent height (positive up). Furthermore, there is a convention to orient the *x*-axis toward the viewer, biased either to the right or left. If a diagram (3D projection or 2D perspective drawing) shows the *x*- and *y*-axis horizontally and vertically, respectively, then the *z*-axis should be shown pointing "out of the page" towards the viewer or camera. In such a 2D diagram of a 3D coordinate system, the *z*-axis would appear as a line or ray pointing down and to the left or down and to the right, depending on the presumed viewer or camera perspective. In any diagram or display, the orientation of the three axes, as a whole, is arbitrary. However, the orientation of the axes relative to each other should always comply with the right-hand rule, unless specifically stated otherwise. All laws of physics and math assume this right-handedness, which ensures consistency.

For 3D diagrams, the names "abscissa" and "ordinate" are rarely used for *x* and *y*, respectively. When they are, the *z*-coordinate is sometimes called the applicate. The words *abscissa*, *ordinate* and *applicate* are sometimes used to refer to coordinate axes rather than the coordinate values.

Quadrants and Octants

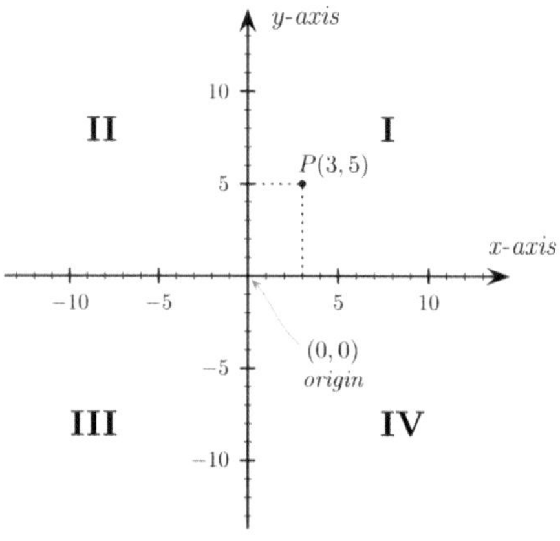

The four quadrants of a Cartesian coordinate system

The axes of a two-dimensional Cartesian system divide the plane into four infinite regions, called quadrants, each bounded by two half-axes. These are often numbered from 1st to 4th and denoted by Roman numerals: I (where the signs of the two coordinates are I (+,+), II (−,+), III (−,−), and IV (+,−). When the axes are drawn according to the mathematical custom, the numbering goes counter-clockwise starting from the upper right ("north-east") quadrant.

Similarly, a three-dimensional Cartesian system defines a division of space into eight regions or octants, according to the signs of the coordinates of the points. The convention used for naming a specific octant is to list its signs, e.g. (+ + +) or (− + −). The generalization of the quadrant and octant to an arbitrary number of dimensions is the orthant, and a similar naming system applies.

Applications

Cartesian coordinates are an abstraction that have a multitude of possible applications in the real world. However, three constructive steps are involved in superimposing coordinates on a problem application. 1) Units of distance must be decided defining the spatial size represented by the numbers used as coordinates. 2) An origin must be assigned to a specific spatial location or landmark, and 3) the orientation of the axes must be defined using available directional cues for all but one axis.

Consider as an example superimposing 3D Cartesian coordinates over all points on the Earth (i.e. geospatial 3D). What units make sense? Kilometers are a good choice, since the original definition of the kilometer was geospatial, 10,000 km equalling the surface distance from the Equator to the North Pole. Where to place the origin? Based on symmetry, the gravitational center of the Earth suggests a natural landmark (which can be sensed via satellite orbits). Finally, how to orient X-, Y- and Z-axis directions? The axis of Earth's spin provides a natural direction strongly associated with "up vs. down", so positive Z can adopt the direction from geocenter to North Pole. A location on the Equator is needed to define the X-axis, and the prime meridian stands out as a reference direction, so the X-axis takes the direction from geocenter out to [0 degrees longitude, 0 degrees latitude]. Note that with 3 dimensions, and two perpendicular axes directions pinned down for X and Z, the Y-axis is determined by the first two choices. In order to obey the right-hand rule, the Y-axis must point out from the geocenter to [90 degrees longitude, 0 degrees latitude]. So what are the geocentric coordinates of the Empire State Building in New York City? Using [longitude = −73.985656, latitude = 40.748433], Earth radius = $40{,}000 / 2\pi$, and transforming from spherical --> Cartesian coordinates, you can estimate the geocentric coordinates of the Empire State Building, [x, y, z] = [1330.53 km, −4635.75 km, 4155.46 km]. GPS navigation relies on such geocentric coordinates.

In engineering projects, agreement on the definition of coordinates is a crucial foundation. One cannot assume that coordinates come predefined for a novel application, so knowledge of how to erect a coordinate system where there is none is essential to applying René Descartes' ingenious thinking.

While spatial apps employ identical units along all axes, in business and scientific apps, each axis may have different units of measurement associated with it (such as kilograms, seconds, pounds, etc.). Although four- and higher-dimensional spaces are difficult to visualize, the algebra of Cartesian coordinates can be extended relatively easily to four or more variables, so that certain calculations involving many variables can be done. (This sort of algebraic extension is what is used to define the geometry of higher-dimensional spaces.) Conversely, it is often helpful to use the geometry of Cartesian coordinates in two or three dimensions to visualize algebraic relationships between two or three of many non-spatial variables.

The graph of a function or relation is the set of all points satisfying that function or relation. For a function of one variable, f, the set of all points (x, y), where $y = f(x)$ is the graph of the function f. For a function g of two variables, the set of all points (x, y, z), where z = g(x, y) is the graph of the function g. A sketch of the graph of such a function or relation would consist of all the salient parts of the function or relation which would include its relative extrema, its concavity and points of inflection, any points of discontinuity and its end behavior. All of these terms are more fully defined in calculus. Such graphs are useful in calculus to understand the nature and behavior of a function or relation.

Geographic Coordinate Conversion

GIS specialists often need to transform data from one coordinate system and/or datum to another. For example, digital data produced by tracing paper maps over a digitizing tablet need to be transformed from the tablet's non-georeferenced plane coordinate system into a georeferenced plane or spherical coordinate system that can be georegistered with other digital data "layers." Raw image data produced by scanning the Earth's surface from space tend to be skewed geometrically as a result of satellite orbits and other factors; to be useful these too need to be transformed into georeferenced coordinate systems. Even the point data produced by GPS receivers, which are measured as latitude and longitude coordinates based upon the WGS84 datum, often need to be transformed to other coordinate systems or datums to match project specifications.

After defining the coordinate system that matches your data, you may still find you want to use data in a different coordinate system. This is where transformations are useful. Transformations are required to convert data between different geographic coordinate systems or between different vertical coordinate systems. Unless your data lines up, you'll face difficulties and inaccuracies in any analysis and mapping you perform on the mismatched data.

Coordinate Format Conversion

Informally, specifying a geographic location usually means giving the location's latitude and longitude. The numerical values for latitude and longitude can occur in a number of different formats:

- degrees minutes seconds : 40° 26′ 46″ N 79° 58′ 56″ W
- degrees decimal minutes: 40° 26.767′ N 79° 58.933′ W
- decimal degrees: 40.446° N 79.982° W

There are 60 minutes in a degree and 60 seconds in a minute. Therefore, to convert from a degrees minutes seconds format to a decimal degrees format, one may use the formula

$$\text{decimal degrees}=\text{degrees}+\text{minutes}/60+\text{seconds}/3600.$$

To convert back from decimal degree format to degrees minutes seconds format,

$$\begin{aligned}\text{degrees} &= \lfloor \text{decimal degrees} \rfloor \\ \text{minutes} &= \lfloor 60 \times (\text{decimal degrees-degrees}) \rfloor \\ \text{seconds} &= 3600 \times (\text{decimal degrees-degrees}) - 60 \times \text{minutes}\end{aligned}$$

where the notation $\lfloor x \rfloor$ means take the integer part of x and is called a floor function.

Coordinate System Conversion

A coordinate system conversion is a conversion from one coordinate system to another, with both coordinate systems based on the same geodetic datum. Common conversion tasks include conversion between geodetic and ECEF coordinates and conversion from one type of map projection to another.

From Geodetic to ECEF Coordinates

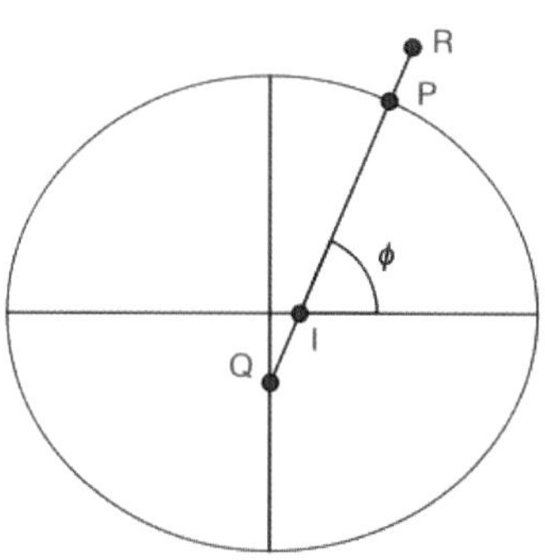

The length PQ is called prime vertical radius $(N(\phi))$. The length IQ is equal to $e^2 N(\phi)$ R $=(X,Y,Z)$.

Geodetic coordinates (latitude ϕ, longitude λ, height h) can be converted into ECEF coordinates using the following equation:

$$X = \left(N(\phi) + h\right)\cos\phi\cos\lambda$$
$$Y = \left(N(\phi) + h\right)\cos\phi\sin\lambda$$
$$Z = \left(\frac{b^2}{a^2}N(\phi) + h\right)\sin\phi$$

where

$$N(\phi) = \frac{a^2}{\sqrt{a^2\cos^2\phi + b^2\sin^2\phi}} = \frac{a}{\sqrt{1 - e^2\sin^2\phi}},$$

and a and b are the equatorial radius (semi-major axis) and the polar radius (semi-minor axis), respectively. $e^2 = 1 - \frac{b^2}{a^2}$ is the square of the first numerical eccentricity of the ellipsoid. The prime vertical radius of curvature $N(\phi)$ is the distance from the surface to the Z-axis along the ellipsoid normal. The parameter h is eliminated by subtracting

$$e^2 = 1 - \frac{b^2}{a^2}a \quad \frac{\sqrt{X^2 + Y^2}}{\cos\phi} = N + h$$

and

$$\frac{Z}{\sin\phi} = \frac{b^2}{a^2}N + h,$$

so the following equation holds:

$$\frac{p}{\cos\phi} - \frac{Z}{\sin\phi} - e^2 N(\phi) = 0,$$

where $p = \sqrt{X^2 + Y^2}$.

The orthogonality of the coordinates is confirmed via differentiation:

$$\begin{pmatrix} dX \\ dY \\ dZ \end{pmatrix} = \begin{pmatrix} -\sin\lambda & -\sin\phi\cos\lambda & \cos\phi\cos\lambda \\ \cos\lambda & -\sin\phi\sin\lambda & \cos\phi\sin\lambda \\ 0 & \cos\phi & \sin\phi \end{pmatrix}\begin{pmatrix} dE \\ dN \\ dU \end{pmatrix},$$

$$\begin{pmatrix} dE \\ dN \\ dU \end{pmatrix} = \begin{pmatrix} \left(N(\phi) + h\right)\cos\phi & 0 & 0 \\ 0 & M(\phi) + h & 0 \\ 0 & 0 & 1 \end{pmatrix}\begin{pmatrix} d\lambda \\ d\phi \\ dh \end{pmatrix},$$

where

$$M(\phi) = \frac{a(1-e^2)}{\left(1-e^2\sin^2\phi\right)^{3/2}}$$

From ECEF to Geodetic Coordinates

The conversion of ECEF coordinates to geodetic coordinates involves more trigonometry but is sensitive to small accuracy due to Rn and h being maybe 10^6 apart, but longitude is same as geocentric, λ.

There are several methods that solve the equation; two are shown.

Newton–Raphson Method

The following Bowring's irrational geodetic-latitude equation is efficient to be solved by Newton–Raphson iteration method:

$$\kappa - 1 - \frac{e^2 a \kappa}{\sqrt{p^2 + (1-e^2)Z^2\kappa^2}} = 0,$$

where $\kappa = \dfrac{p}{Z}\tan\phi$ and $p = \sqrt{x^2+y^2}$. The height is calculated as:

$$h = e^{-2}(\kappa^{-1} - \kappa_0^{-1})\sqrt{p^2 + Z^2\kappa^2},$$

$$\kappa_0 = \left(1-e^2\right)^{-1}.$$

The iteration can be transformed into the following calculation:

$$\kappa_{i+1} = \frac{c_i + \left(1-e^2\right)Z^2\kappa_i^3}{c_i - p^2} = 1 + \frac{p^2 + \left(1-e^2\right)Z^2\kappa_i^3}{c_i - p^2},$$

where $c_i = \dfrac{\left(p^2 + \left(1-e^2\right)Z^2\kappa_i^2\right)^{3/2}}{ae^2}$.

The constant κ_0 is a good starter value for the iteration when $h \approx 0$. Bowring showed that the single iteration produces a sufficiently accurate solution. He used extra trigonometric functions in his original formulation.

Ferrari's Solution

The quartic equation of κ, derived from the above, can be solved by Ferrari's solution to yield:

$$\zeta = (1 - e^2)z^2 / a^2,$$
$$\rho = (p^2 / a^2 + \zeta - e^4) / 6,$$
$$s = e^4 \zeta p^2 / (4a^2),$$
$$t = \sqrt[3]{\rho^3 + s + \sqrt{s(s + 2\rho^3)}},$$
$$u = \rho + t + \rho^2 / t,$$
$$v \quad \sqrt{u^2 + e^4\zeta},$$
$$w = e^2(u + v - \zeta) / (2v),$$
$$\kappa = 1 + e^2(\sqrt{u + v + w^2} + w) / (u + v).$$

The Application of Ferrari's Solution

A number of techniques and algorithms are available but the most accurate according to Zhu, is the following 15 step procedure summarised by Kaplan. It is assumed that geodetic parameters $\{a, b, e, e'\}$ are known

$$r = \sqrt{X^2 + Y^2}$$
$$e'^2 = (a^2 - b^2) / b^2$$
$$E^2 = a^2 - b^2$$
$$F = 54b^2 Z^2$$
$$G = r^2 + (1 - e^2)Z^2 - e^2 E^2$$
$$C = \frac{e^4 F r^2}{G^3}$$
$$S = \sqrt[3]{1 + C + \sqrt{C^2 + 2C}}$$
$$P = \frac{F}{3\left(S + \frac{1}{S} + 1\right)^2 G^2}$$
$$Q = \sqrt{1 + 2e^4 P}$$
$$r_0 = \frac{-(Pe^2 r)}{1 + Q} + \sqrt{\frac{1}{2}a^2(1 + 1/Q) - \frac{P(1 - e^2)Z^2}{Q(1 + Q)} - \frac{1}{2}Pr^2}$$
$$U = \sqrt{(r - e^2 r_0)^2 + Z^2}$$
$$V = \sqrt{(r - e^2 r_0)^2 + (1 - e^2)Z^2}$$
$$Z_0 = \frac{b^2 Z}{aV}$$
$$h = U\left(1 - \frac{b^2}{aV}\right)$$
$$\phi = \arctan\left[\frac{Z + e'^2 Z_0}{r}\right]$$
$$\lambda = \arctan 2[Y, X]$$

Note: arctan2[Y,X] is the four-quadrant inverse tangent function.

Power Series

For small e^2 the power series

$$\kappa = \sum_{i \geq 0} \alpha_i e^{2i}$$

starts with

$$\alpha_0 = 1;$$

$$\alpha_1 = \frac{a}{\sqrt{Z^2 + p^2}};$$

$$\alpha_2 = \frac{aZ^2\sqrt{Z^2 + p^2} + 2a^2p^2}{2(Z^2 + p^2)^2}.$$

Geodetic to/from ENU Coordinates

To convert from geodetic coordinates to local ENU coordinates is a two-stage process:

1. Convert geodetic coordinates to ECEF coordinates
2. Convert ECEF coordinates to local ENU coordinates

From ECEF to ENU

To transform from ECEF coordinates to the local coordinates we need a local reference point, typically this might be the location of a radar. If a radar is located at $\{X_r, Y_r, Z_r\}$ and an aircraft at $\{X_p, Y_p, Z_p\}$ then the vector pointing from the radar to the aircraft in the ENU frame is

$$\begin{bmatrix} x \\ y \\ z \end{bmatrix} = \begin{bmatrix} -\sin\lambda_r & \cos\lambda_r & 0 \\ -\sin\phi_r \cos\lambda_r & -\sin\phi_r \sin\lambda_r & \cos\phi_r \\ \cos\phi_r \cos\lambda_r & \cos\phi_r \sin\lambda_r & \sin\phi_r \end{bmatrix} \begin{bmatrix} X_p - X_r \\ Y_p - Y_r \\ Z_p - Z_r \end{bmatrix}$$

Note: ϕ is the geodetic latitude. A prior version of this page showed use of the geocentric latitude (ϕ'). The geocentric latitude is not the appropriate up direction for the local tangent plane. If the original geodetic latitude is available it should be used, otherwise, the relationship between geodetic and geocentric latitude has an altitude dependency, and is captured by:

$$\tan\phi' = \frac{Z_r}{\sqrt{X_r^2 + Y_r^2}} = \frac{N(\phi)(1-f)^2 + h}{N(\phi) + h}\tan\phi$$

Obtaining *geodetic* latitude from *geocentric* coordinates from this relationship requires an iterative solution approach, otherwise the *geodetic* coordinates may be computed via the approach in the section above labeled "From ECEF to geodetic coordinates."

The geocentric and geodetic longitude have the same value. This is true for the Earth and other similar shaped planets because their latitude lines (parallels) can be considered in much more degree perfect circles when compared to their longitude lines (meridians).

$$\tan\lambda = \frac{Y_r}{X_r}$$

Note: Unambiguous determination of ϕ and λ requires knowledge of which quadrant the coordinates lie in.

From ENU to ECEF

This is just the inversion of the ECEF to ENU transformation so

$$\begin{bmatrix} X \\ Y \\ Z \end{bmatrix} = \begin{bmatrix} -\sin\lambda & -\sin\phi\cos\lambda & \cos\phi\cos\lambda \\ \lambda\ \cos\lambda & -\sin\phi\sin\lambda & \cos\phi\sin\lambda \\ 0 & \cos\phi & \sin\phi \end{bmatrix} \begin{bmatrix} x \\ y \\ z \end{bmatrix} + \begin{bmatrix} X_r \\ Y_r \\ Z_r \end{bmatrix}$$

Conversion Across Map Projections

Conversion of coordinates and map positions among different map projections reference to the same datum may be accomplished either through direct translation formulas from one projection to another, or by first converting from a projection A to an intermediate coordinate system, such as ECEF, then converting from ECEF to projection B. The formulas involved can be complex and in some cases, such as in the ECEF to geodetic conversion above, the conversion has no closed-form solution and approximate methods must be used. References such as the DMA Technical Manual 8358.1 and the USGS paper Map Projections: A Working Manual contain formulas for conversion of map projections. It is common to use computer programs to perform coordinate conversion tasks, such as with the DoD and NGA supported GEOTRANS program.

Datum Transformations

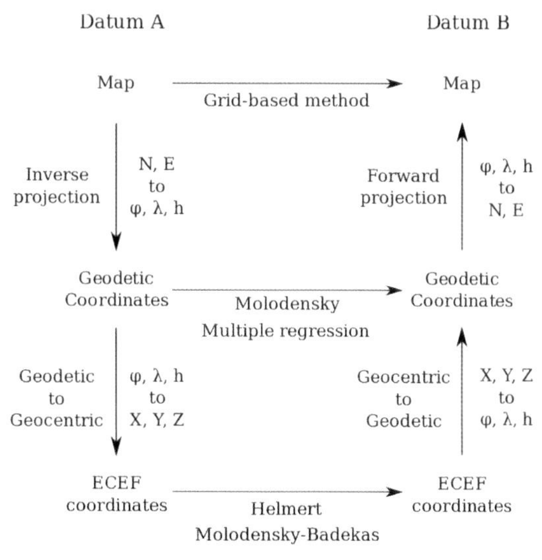

Transformations among datums can be accomplished in a number of ways. There are transformations that directly convert geodetic coordinates from one datum to another. There are more indirect transforms that convert from geodetic coordinates to ECEF coordinates, transform the ECEF coordinates from one datum to the another, then transform ECEF coordinates of the new datum back to geodetic coordinates. There are also grid-based transformations that directly transform from one (datum, map projection) pair to another (datum, map projection) pair.

Grid-based Method

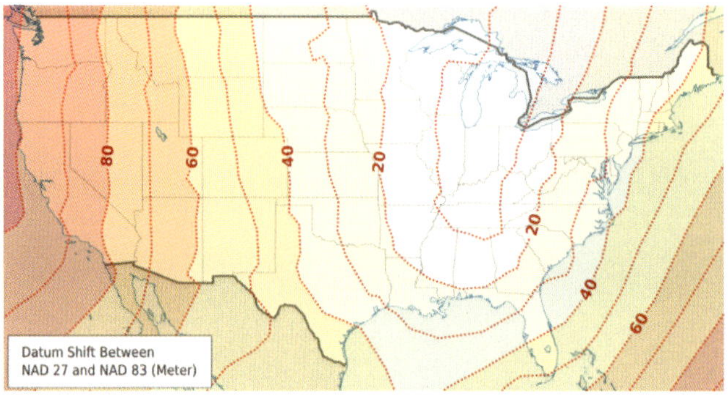

Magnitude of shift in position between NAD27 and NAD83 datum as a function of location.

Grid-based transformations directly convert map coordinates from one (map-projection, geodetic datum) pair to map coordinates of another (map-projection, geodetic datum) pair. An example is the NADCON method for transforming from the North American Datum (NAD) 1927 to the NAD 1983 datum. The High Accuracy Reference Network (HARN), a high accuracy version of the NADCON transforms, have an accuracy of approximately 5 centimeters. The National Transformation version 2 (NTv2) is a Canadian version of NADCON for transforming between NAD 1927 and NAD 1983. HARNs are also known as NAD 83/91 and High Precision Grid Networks (HPGN). Subsequently, Australia and New Zealand adopted the NTv2 format to create grid-based methods for transforming among their own local datums.

Like the multiple regression equation transform, grid-based methods use a low-order interpolation method for converting map coordinates, but in two dimensions instead of three. The NOAA provides a software tool (as part of the NGS Geodetic Toolkit) for performing NADCON transformations.

Molodensky transformation

The Molodensky transformation converts directly between geodetic coordinate systems of different datums without the intermediate step of converting to geocentric ECEF coordinates. It requires the three shifts between the datum centers and the differences between the reference ellipsoid semi-major axes and flattening parameters.

The Molodensky transform is used by the National Geospatial-Intelligence Agency (NGA) in their standard TR8350.2 and the NGA supported GEOTRANS program. The Molodensky method was popular before the advent of modern computers and the method is part of many geodetic programs.

Multiple Regression Equations

Datum transformations through the use of empirical multiple regression methods were created to achieve higher accuracy results over small geographic regions than the standard Molodensky transformations. MRE transforms are used to transform local datums over continent-sized or smaller regions to global datums, such as WGS 84. The standard NIMA TM 8350.2, Appendix D, lists MRE transforms from several local datums to WGS 84, with accuracies of about 2 meters.

The MREs are a direct transformation of geodetic coordinates with no intermediate ECEF step. Geodetic coordinates ϕ_B, λ_B, h_B in the new datum B are modeled as polynomials of up to the ninth degree in the geodetic coordinates ϕ_A, λ_A, h_A of the original datum A. For instance, the change in ϕ_B could be parameterized as (with only up to quadratic terms shown):9

$$\Delta\phi = a_0 + a_1 U + a_2 V + a_3 U^2 + a_4 UV + a_5 V^2 +$$

where

$$\Delta\phi = a_0 + a_1 U + a_2 V + a_3 U^2 + a_4 UV + a_5 V^2 +$$

$$a_i = \text{parameters fitted by multiple regression}$$

$$U = K(\phi_A - \phi_m)$$

$$V = K(\lambda_A - \lambda_m)$$

$$K = \text{scale factor}$$

$$\phi_m, \lambda_m = \text{origin of the datum A}$$

with similar equations for $\Delta\lambda$ and Δh. Given a sufficient number of (A,B) coordinate pairs for landmarks in both datums for good statistics, multiple regression methods are used to fit the parameters of these polynomials. The polynomials, along with the fitted coefficients, form the multiple regression equations.

Helmert Transformation

Use of the Helmert transform in the transformation from geodetic coordinates of datum A to geodetic coordinates of datum B occurs in the context of a three-step process:

1. Convert from geodetic coordinates to ECEF coordinates for datum A
2. Apply the Helmert transform, with the appropriate $A \to B$ transform parameters, to transform from datum A ECEF coordinates to datum B ECEF coordinates
3. Convert from ECEF coordinates to geodetic coordinates for datum B

In terms of ECEF XYZ vectors, the Helmert transform has the form

$$\begin{bmatrix} X_B \\ Y_B \\ Z_B \end{bmatrix} = \begin{bmatrix} c_x \\ c_y \\ c_z \end{bmatrix} + (1 + s \times 10^{-6}) \cdot \begin{bmatrix} 1 & -r_z & r_y \\ r_z & 1 & -r_x \\ -r_y & r_x & 1 \end{bmatrix} \cdot \begin{bmatrix} X_A \\ Y_A \\ Z_A \end{bmatrix}.$$

The Helmert transform is a seven-parameter transform with three translation (shift) parameters c_x, c_y, c_z, three rotation parameters r_x, r_y, r_z and one scaling (dilation) parameter s. The Helmert transform is an approximate method that is accurate when the transform parameters are small relative to the magnitudes of the ECEF vectors. Under these conditions, the transform is considered reversible.

A fourteen-parameter Helmert transform, with linear time dependence for each parameter, can be used to capture the time evolution of geographic coordinates dues to geomorphic processes, such as continental drift. and earthquakes. This has been incorporated into software, such as the Horizontal Time Dependent Positioning (HTDP) tool from the U.S. NGS.

Molodensky-badekas Transformation

To eliminate the coupling between the rotations and translations of the Helmert transform, three additional parameters can be introduced to give a new XYZ center of rotation closer to coordinates being transformed. This ten-parameter model is called the *Molodensky-Badekas transformation* and should not be confused with the more basic Molodensky transform.

Like the Helmert transform, using the Molodensky-Badekas transform is a three-step process:

1. Convert from geodetic coordinates to ECEF coordinates for datum A
2. Apply the Molodensky-Badekas transform, with the appropriate $A \to B$ transform parameters, to transform from datum A ECEF coordinates to datum B ECEF coordinates
3. Convert from ECEF coordinates to geodetic coordinates for datum

The transform has the form

$$\begin{bmatrix} X_B \\ Y_B \\ Z_B \end{bmatrix} = \begin{bmatrix} X_A \\ Y_A \\ Z_A \end{bmatrix} + \begin{bmatrix} \Delta X_A \\ \Delta Y_A \\ \Delta Z_A \end{bmatrix} + \begin{bmatrix} 1 & -r_z & r_y \\ r_z & 1 & -r_x \\ -r_y & r_x & 1 \end{bmatrix} \cdot \begin{bmatrix} X_A - X_A^0 \\ Y_A - Y_A^0 \\ Z_A - Z_A^0 \end{bmatrix} + \Delta S \begin{bmatrix} X_A - X_A^0 \\ Y_A - Y_A^0 \\ Z_A - Z_A^0 \end{bmatrix}.$$

where (X_A^0, Y_A^0, Z_A^0) is the origin for the rotation and scaling transforms and ΔS is the scaling factor.

The Molodensky-Badekas transform is used to transform local geodetic datums to a global geodetic datum, such as WGS 84. Unlike the Helmert transform, the Molodensky-Badekas transform is not reversible due to the rotational origin being associated with the original datum.

Linear Referencing

Linear referencing is the method of storing geographic locations by using relative positions along a measured linear feature. Distance measures are used to locate events along the line:

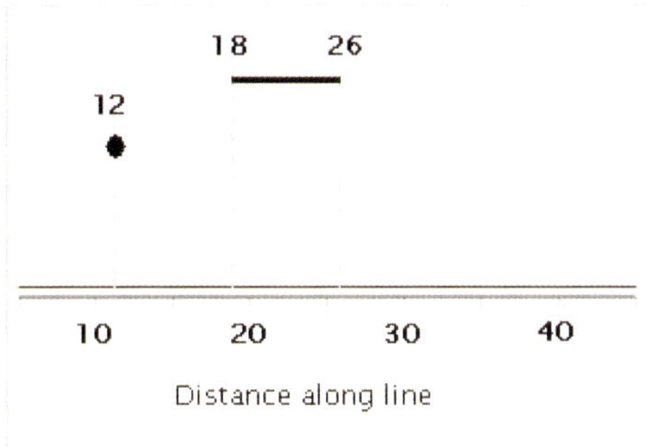

Measurements along features are used to locate point events and line events using a number of conventions. Here are some common examples:

A point can be located along the line in the graph below:

- At measure 12 along the line
- 4 units east of measure marker 10 along the line

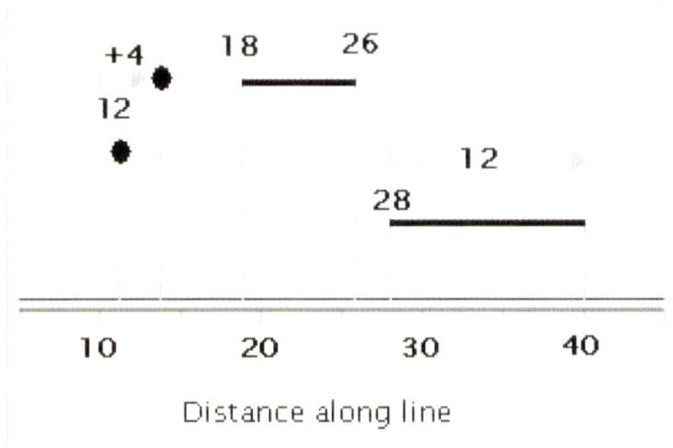

Line features can be referenced in a few ways. In this example above

- The line starts at measurement 18 and ends at 26.
- The line starts at measurement 28 and continues for 12 units.

Why use Linear Referencing?

Linear referencing is used for many reasons. The following are the two primary reasons:

- Many locations are recorded as events along linear features—for example, locations of traffic accidents are recorded using a convention such as "27 meters east of reference mile marker 35 along State Highway 287." Many sensors record conditions along linear features using measures of distance or time along the lines—along pipelines, along roads, along streams, and so forth.
- Linear referencing is also used to associate multiple sets of attributes to portions of linear features without requiring that underlying lines be segmented (split) each time that attribute values change. For example, most road centerline feature classes are segmented where three or more road segments intersect and where the road names change.

Users often want to record many additional attributes about the roads. Without the use of linear referencing, this could require that roads be split into many tiny segments at each location where attribute values change. As an alternative, these situations can be handled as linear referencing events along the roads as in the figure below:

What is Dynamic Segmentation?

Dynamic segmentation is the process of computing the map locations of events stored and managed in an event table using a linear referencing measurement system and displaying them on a map. The term dynamic segmentation is derived from the concept that line features need not be split (in other words, "segmented") each time an attribute value changes; you can "dynamically" locate the segment.

Using dynamic segmentation, multiple sets of attributes can be associated with any portion of an existing linear feature independently of where it begins or ends. These attributes can be displayed, queried, edited, and analyzed without affecting the underlying linear feature's geometry.

In most data models, linear features are split at intersections where they connect with two or more line features and frequently are also split where key attribute values change, such as a change in the road name.

However, users often want to record many additional attributes about the roads. This could require that roads be split into many tiny segments at each location where attribute values change. As an alternative, these situations can be handled as linear referencing events along the roads as illustrated in the concept figure below:

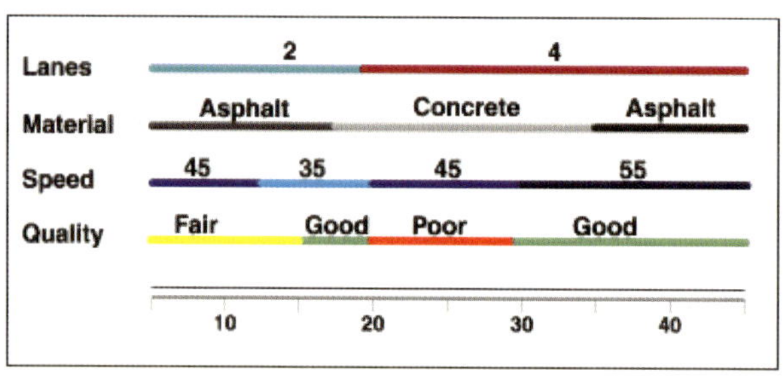

Certain linear features also have attributes that change frequently—for example, observations that describe the condition of segments along key infrastructure networks (roads, pipes, and so on). Plus, these observations could be made repeatedly over time. For example, the pavement condition of a road changes as pavement deteriorates and is subsequently repaired. Linear referencing can be used to accurately locate multiple observations of the pavement condition, as illustrated below.

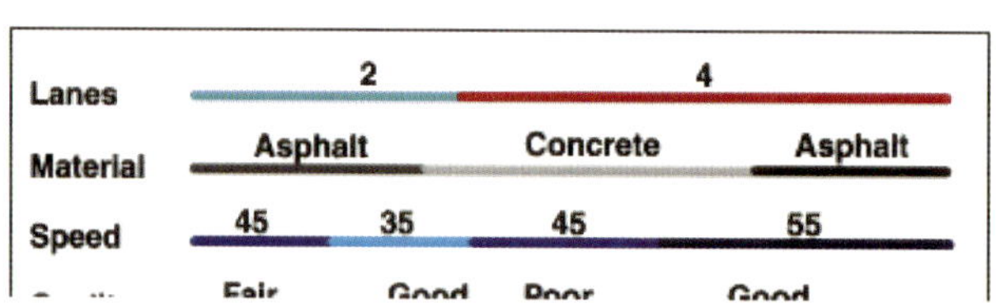

Segmenting linear features on a frequent basis becomes even more problematic when you consider that you may need to store other attributes. In addition to road conditions, you might also want to record traffic volumes, lane information, surface material, speed limits, and accident locations.

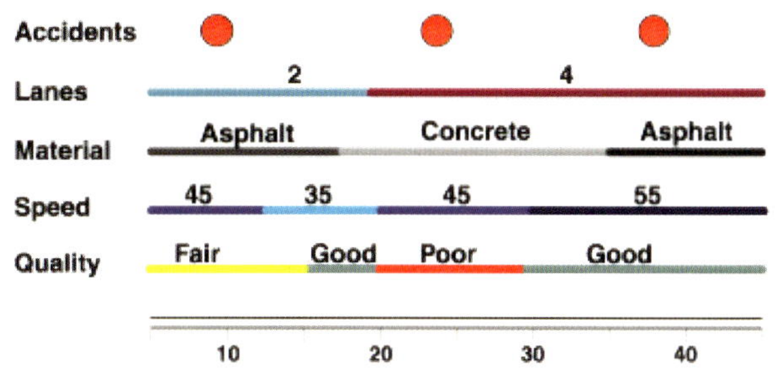

Applications

Highways and Streets

Agencies that manage highways and streets use linear referencing in a variety of ways in their day-to-day operations. For example, linear referencing is useful for the following:

- Assessing pavement conditions
- Maintaining, managing, and valuing assets—for example, traffic signs and signals, guard rails, toll booths, and loop detectors
- Organizing bridge management information
- Reviewing and coordinating construction projects

Linear referencing also facilitates the creation of a common database that traffic planners, traffic engineers, and public works analysts can use for cross-disciplinary decision support.

The following is an example displaying pavement conditions.

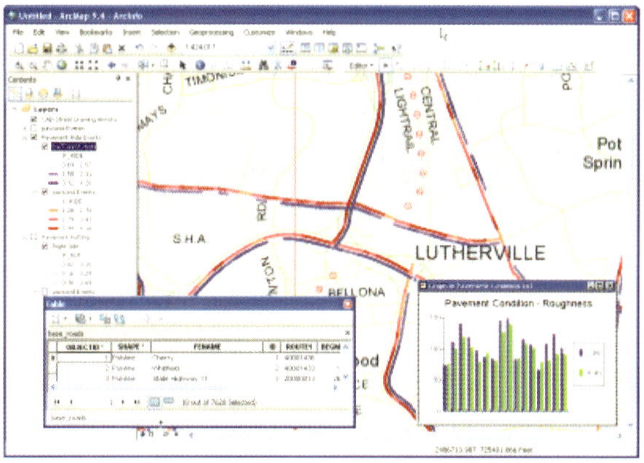

Pavement conditions

Transit

Linear referencing is a key component in transit applications, and it facilitates such activities as these:

- Route planning and analysis
- Automatic vehicle location and tracking

- Bus stop and facility inventory
- Rail system facility management
- Track, power, communications, and signal maintenance
- Accident reporting and analysis
- Demographic analysis and route restructuring
- Ridership analysis and reporting
- Transportation planning and modeling

The following shows the results of a corridor study, displaying the number of traffic accidents along a stretch of highway.

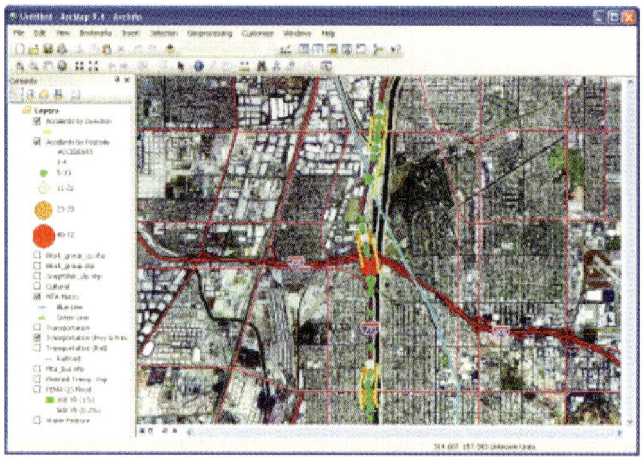

Traffic accidents in a corridor study

Railways

Railways use linear referencing to manage key information for rail operations, maintenance, asset management, and decision support systems. Linear referencing makes it possible, for example, to select a line and track and identify milepost locations for bridges and other obstructions that would prevent various types of freight movement along the route. Furthermore, linear referencing can be used to display track characteristics or view digital images of bridges and obstructions.

The following is an example of analyzing rail clearances along a rail line.

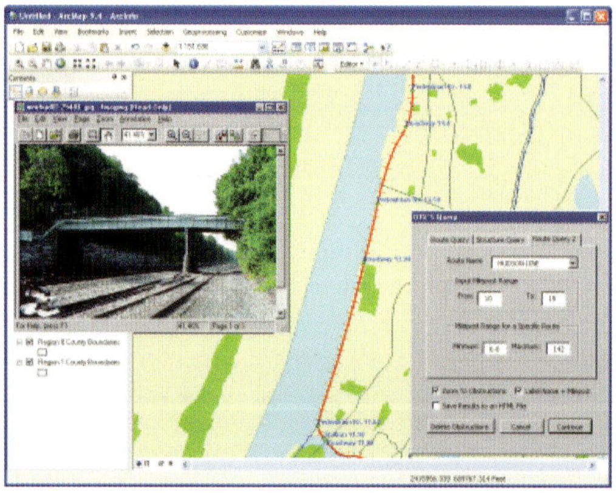

Analyzing rail line clearance

Oil and Gas Exploration

The petroleum industry manages tremendous volumes of data used in geophysical exploration. Seismic surveys, or shotpoint data, are used to help understand the underlying geology in an area. The nature of seismic data is that it must be represented as both a linear object—the seismic line—and a collection of point objects (the shotpoints). Both the seismic line and the individual shotpoints have attributes, must be maintained at the same time, and are used in modeling applications. Linear referencing helps solve this problem.

The following shows an example of posting and labeling seismic lines and shotpoints in the Gulf of Mexico.

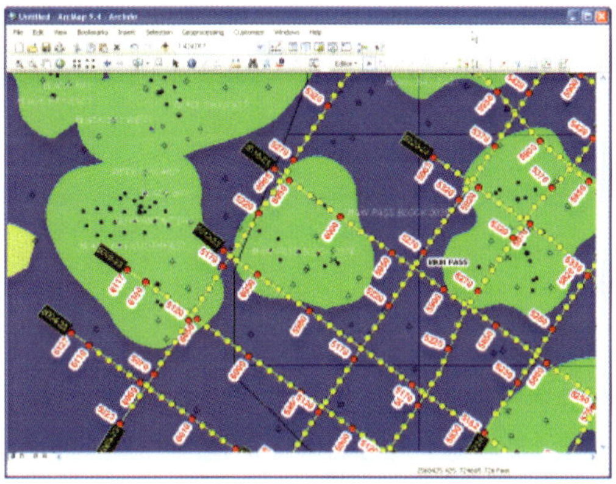

Posting and labeling seismic lines and shotpoints

Pipelines

In the pipeline industry, linear referencing is often referred to as stationing. Stationing allows any point along a pipeline to be uniquely identified. As such, stationing is useful in these applications:

In the following image, pipeline coating material types are being examined.

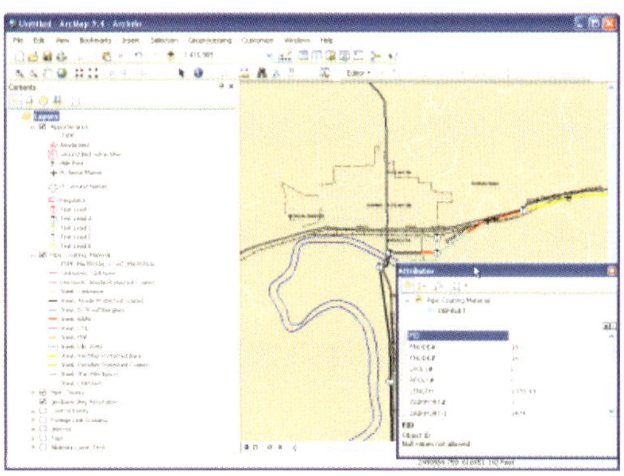

Pipeline coating material types are being examined.

- Collecting and storing information regarding pipeline facilities
- Inline and physical inspection histories
- Regulatory compliance information

- Risk assessment studies
- Work history events
- Geographic information, such as environmentally sensitive areas, political boundaries (for example, state and county), right-of-way boundaries, and various types of crossings

Water Resources

In hydrology applications, linear referencing is often called river addressing. River addressing allows objects such as field monitoring stations—which collect information about water quality analysis, toxic release inventories, drinking water supplies, flow, and so on—to be located along a river or stream system. Furthermore, the measurement scheme used in river addressing allows the measurement of flow distance between any two points on a flow path.

In the image below, monitoring stations along a hydrology network are identified.

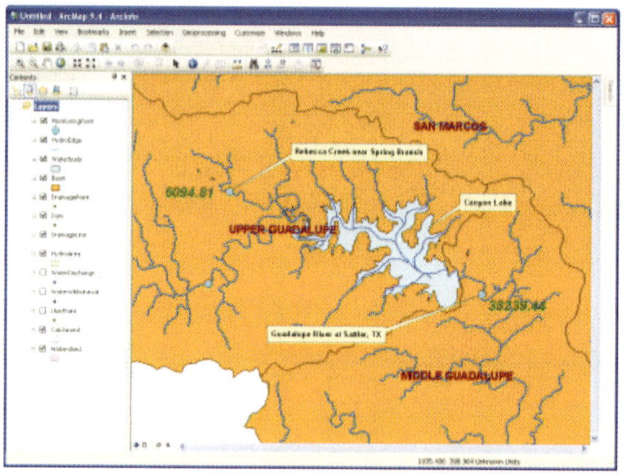

Identifying monitoring stations along a hydrology network

Satellite Navigation

Using the fixed stars, you can determine your position anywhere on the Earth's surface to within a few hundred metres, provided the weather is fine. But navigation satellites can tell you where you are to the nearest few metres or better, whatever the weather. Enhanced instruments can even pinpoint the position of a stationary object to within a few centimetres by measuring the object's position many thousands of times over several hours and then working out the average of the measurements.

All you need to take advantage of this sophisticated new technology is a receiver to pick up signals transmitted by navigational satellites. The receiver technology is small enough to be incorporated into the electronics of a car or mobile phone, and all our lives are changing as a result.

How does it Work?

To pinpoint your location accurately, your receiver needs to receive signals from at least four navigational satellites. The receiver determines your distance from each of the satellites by measuring the time taken by the signal to travel from the satellite to your receiver antenna.

Distance measurements from two satellites tell you that you are situated somewhere on the circle where two spheres intersect. The spheres each have one of the two satellites at their centre and their radii are the satellite-receiver distances. Knowing your distance from a third satellite fixes your position at one of the two points where the circle intersects the third sphere. One of the intersection points can usually be discounted – for instance, it may be thousands of kilometres above the Earth's surface.

In practice, a fourth satellite is needed to synchronise your receiver's clock with a common time standard which is strictly adhered to by the clocks on board all the satellites. The use of a fourth satellite also resolves the position ambiguity that occurs with only three satellites. In general, the more satellites used, the greater the positioning accuracy. Many receivers have channels for receiving signals from up to 15 satellites.

The Importance of Measuring Time

Determining precise location depends on accurately measuring the distances between receiver and satellite, and that depends on very accurate measurement of the radio signal's travel time from the satellite to the receiver. As these signals travel at the speed of light, the journey times are tiny fractions of a second.

The receiver measures travel times by comparing 'time marks' imprinted on the satellite signals with the time recorded on the receiver's clock. The time marks are controlled by a highly accurate atomic clock on board each satellite.

These clocks, however, are too expensive to incorporate into standard receivers, which have to make do with small quartz oscillators like those found in a wristwatch. Quartz oscillators are very accurate when measuring times of less than a few seconds, but rather inaccurate over longer periods. The solution is to re-set the receiver's time to the satellite's time continuously. This is done by the receiver's processor using an approximation method involving signals from at least four satellites.

For this system of measurement to work, all satellites need to be synchronised so that they can start transmitting their signals at precisely the same time. This is achieved by continuously synchronising all on-board atomic clocks with a master clock on the ground. These super-accurate clocks have an accuracy equivalent to one second in three million years.

Where are the Navigational Satellites?

It is only possible to determine a location on Earth if you know the location of the navigational satellites very precisely. This is achieved by placing the satellites in highly stable Medium Earth Orbits (MEOs) at an altitude of about 22 000 kilometres. MEOs are the orbits of choice for a number of reasons: their stability enables exact orbit predictions; the satellites travel relatively slowly and so can be observed over several hours, and the satellites can be arranged in a constellation so that at least four are visible from any point on the Earth's surface at any time.

Despite the predictability of the MEOs, it is still necessary to monitor the precise location of each satellite constantly to achieve high positioning accuracy. This is done from a global network of reference stations on the ground, whose positions are known to within centimetres. The ground system sends data derived from the reference station measurements to each satellite. This precise location information, called ephemeris data, is then relayed by the satellite to the user receivers with the navigation signal.

What about Errors?

Numerous errors can degrade the accuracy of a position measurement. For example, errors in satellite to receiver distances can creep in if conditions within the ionosphere, the electrically charged outer layer of the atmosphere, slow down the signal. Conditions within the ionosphere are influenced by the level of activity on the surface of the Sun. Inaccurate distance measurements will also occur if the signal takes an abnormally long path because it is reflected off tall buildings or other surfaces before reaching the receiver.

There are various ways of overcoming such inaccuracies. The best known is called differential satellite navigation, which uses a fixed receiver in a known position as a reference.

The time taken for the signal to travel from the satellite to the fixed receiver can be calculated precisely because the positions of the fixed receiver and the satellites (and hence the length of the travel path) are known precisely. Any difference between the calculated travel time and that actually measured reflects inaccuracies introduced by disturbances in the ionosphere.

If a moving receiver, attached to an aircraft for example, is within a few hundred kilometres of the fixed receiver, then it is fair to assume that the errors experienced by the signal in reaching both receivers will be roughly the same, as variations in ionospheric conditions tend to be similar over large areas. The timing errors determined by the fixed receiver can then be used to eliminate similar errors in the moving receiver. Major users of satellite navigation, such as large airports, may decide to use the differential technique by installing their own fixed receivers.

Another technique, which makes use of two positioning signals at two different frequencies, does away with the need for differential satellite navigation. It works on the principle that each frequency will be slowed down by a slightly different amount when travelling through the atmosphere. By sending the two frequencies at the same time and recording the time difference between their arrivals, it is possible to build up a model of ionospheric conditions. This dual-frequency technique can increase positioning accuracy to better than one to two metres and will be standard for Galileo.

Finally, it is possible to determine positions to within a few tens of centimetres using a technique called Three Carrier Ambiguity Resolution (TCAR). TCAR eliminates inaccuracies by looking at the wavelike structure of the signal and determining how much it has been pushed along during its passage through the atmosphere. Such phase shifts can be measured to within a fraction of the wavelength, which is typically 20 centimetres.

The calculations needed to do TCAR are presently beyond the capabilities of standard hand-held receivers and this technique is used mainly for long-term monitoring of movement in buildings, oil pipelines, changes in sea level and even changes in the shape of the Earth's crust.

Classification

Satellite navigation systems that provide enhanced accuracy and integrity monitoring usable for civil navigation are classified as follows:

- GNSS-1 is the first generation system and is the combination of existing satellite navigation systems (GPS and GLONASS), with Satellite Based Augmentation Systems (SBAS) or Ground Based Augmentation Systems (GBAS). In the United States, the satellite based component is the Wide Area Augmentation System (WAAS), in Europe it is the European Geostationary Navigation Overlay Service (EGNOS), and in Japan it is the Multi-Functional Satellite Augmentation System (MSAS). Ground based augmentation is provided by systems like the Local Area Augmentation System (LAAS).
- GNSS-2 is the second generation of systems that independently provides a full civilian satellite navigation system, exemplified by the European Galileo positioning system. These systems will provide the accuracy and integrity monitoring necessary for civil navigation; including aircraft. This system consists of L1 and L2 frequencies (in the L band of the radio spectrum) for civil use and L5 for system integrity. Development is also in progress to provide GPS with civil use L2 and L5 frequencies, making it a GNSS-2 system.
- Core Satellite navigation systems, currently GPS (United States), GLONASS (Russian Federation), Galileo (European Union) and Compass (China).
- Global Satellite Based Augmentation Systems (SBAS) such as Omnistar and StarFire.
- Regional SBAS including WAAS (US), EGNOS (EU), MSAS (Japan) and GAGAN (India).
- Regional Satellite Navigation Systems such as China's Beidou, India's NAVIC, and Japan's proposed QZSS.
- Continental scale Ground Based Augmentation Systems (GBAS) for example the Australian GRAS and the joint US Coast Guard, Canadian Coast Guard, US Army Corps of Engineers and US Department of Transportation National Differential GPS (DGPS) service.
- Regional scale GBAS such as CORS networks.
- Local GBAS typified by a single GPS reference station operating Real Time Kinematic (RTK) corrections.

Civil and Military Uses

The original motivation for satellite navigation was for military applications. Satellite navigation allows precision in the delivery of weapons to targets, greatly increasing their lethality whilst reducing inadvertent casualties from mis-directed weapons. Satellite navigation also allows forces to be directed and to locate themselves more easily, reducing the fog of war.

The ability to supply satellite navigation signals is also the ability to deny their availability. The operator of a satellite navigation system potentially has the ability to degrade or eliminate satellite navigation services over any territory it desires.

Satellite navigation using a laptop and a GPS receiver

Augmentation

GNSS augmentation is a method of improving a navigation system's attributes, such as accuracy, reliability, and availability, through the integration of external information into the calculation process, for example, the Wide Area Augmentation System, the European Geostationary Navigation Overlay Service, the Multi-functional Satellite Augmentation System, Differential GPS, GPS Aided GEO Augmented Navigation (GAGAN) and inertial navigation systems.

DORIS

Doppler Orbitography and Radio-positioning Integrated by Satellite (DORIS) is a French precision navigation system. Unlike other GNSS systems, it is based on static emitting stations around the world, the receivers being on satellites, in order to precisely determine their orbital position. The system may be used also for mobile receivers on land with more limited usage and coverage. Used with traditional GNSS systems, it pushes the accuracy of positions to centimetric precision (and to millimetric precision for altimetric application and also allows monitoring very tiny seasonal changes of Earth rotation and deformations), in order to build a much more precise geodesic reference system.

Low Earth Orbit Satellite Phone Networks

The two current operational low Earth orbit satellite phone networks are able to track transceiver units with accuracy of a few kilometers using doppler shift calculations from the satellite. The coordinates are sent back to the transceiver unit where they can be read using AT commands or a graphical user interface. This can also be used by the gateway to enforce restrictions on geographically bound calling plans.

Global Navigation Satellite Systems

Global Navigation Satellite Systems (GNSS) include constellations of Earth-orbiting satellites that broadcast their locations in space and time, of networks of ground control stations, and of receivers that calculate ground positions by trilateration. GNSS are used in all forms of transportation: space stations, aviation, maritime, rail, road and mass transit. Positioning, navigation and timing play a critical role in telecommunications, land surveying, law enforcement, emergency response, precision agriculture, mining, finance, scientific research and so on. They are used to control computer networks, air traffic, power grids and more. Thus the specific objectives of the implementation of the GNSS education curriculum are the demonstration and understanding of GNSS signals, codes, biases and practical applications, and the implications of prospective modernization.

For developing countries, GNSS applications offer a cost-effective way of pursuing sustainable economic growth while protecting the environment. Satellite navigation and positioning data are now used in a wide range of areas that include mapping and surveying, monitoring of the environment, precision agriculture and natural resources management, disaster warning and emergency response, aviation, maritime and land transportation and research areas such as climate change and ionospheric studies.

The successful completion of the work of the International Committee on Global Navigation Systems (ICG), particularly in establishing interoperability among the global systems, will allow a GNSS user to utilize one instrument to receive signals from multiple systems of satellites. This will provide additional data, particularly in urban and mountainous regions, and greater accuracy in timing or position measurements. To benefit from these achievements, GNSS users need to stay abreast of the latest developments in GNSS-related areas and build the capacity to use the GNSS signal.

The performance of GNSS is assessed using four criteria:

1. Accuracy: the difference between a receiver's measured and real position, speed or time;
2. Integrity: a system's capacity to provide a threshold of confidence and, in the event of an anomaly in the positioning data, an alarm;
3. Continuity: a system's ability to function without interruption;
4. Availability: the percentage of time a signal fulfils the above accuracy, integrity and continuity criteria.

Working of GNSS Receivers

Navigation satellites work on the principle of trilateration. Position of an object is determined by its latitude, longitude on the spheroid and height above Mean Sea Level. If at the time of measurement, the instantaneous position of three satellites are known and the distance of the point of measurement from each of these three satellites is known, then the latitude, longitude and height of the point can be determined using simple distance formula. In practice, a fourth satellite is needed to adjust for timing biases.

How does a GPS receiver determine its distance from each satellite? It does so by comparing a code generated by a satellite with the same code generated internally in the receiver. The time difference between the two codes multiplied by the speed of light gives the distance. That requires a very stable signal source on the satellites which is provided by the Rubidium clock which is used to generate the code and the carrier signal for the code. Each satellite has a unique code hence the receiver can identify each satellite in its view. The carrier also contains the precise orbital parameters of the satellite which is updated regularly. The orbital parameters are updated using Integrity and Range Monitoring Stations. A central control station uploads the orbital parameters and synchronizes the Rubidium clocks regularly on each GNSS satellite.

Signals from the GNSS satellites are transmitted in the L-band in five designated frequencies named L1 to L5. Of these L1 and L2 are used by all GNSS satellites, L3 is set apart for transmission from satellites to ground stations to detect nuclear explosions. L4 is to be used for ionospheric corrections and L5 will eventually support safety-of-life applications for aviation and provide improved availability and accuracy.

Global Systems

Global Positioning System, GPS – US

This is the oldest system which began operations in 1978 and grew to 32 planned satellites in Medium Earth Orbit, MEO of 20180 km. The orbits are so selected that at any given time at least six satellites are in view an observer anywhere on earth. Initially exclusive for military use, it was opened to the public in the mid 80s with a selective availability that restricted the positional accuracy to 100m. The military continues to enjoy 5m or better accuracy. SA was discontinued from 2000 during which time the military also developed technologies to deny GPS service on regional basis to potential adversaries. It is widely believed that this was used during the Kargil War.

Glonass – Russia

Starting with a decision in 1978 GLONASS became operational by 1993 over Russia with 12 satellites in 2 orbits at 19,130 km. Its full capacity and global coverage was achieved by 2015 with 23 of 27 satellites operational. The accuracies are comparable to but a little lower than GPS. However, the accuracies are better than GPS in the higher latitudes. The frequencies are the same L1 and L2 but originally adopted a different access mode, FDMA before shifting to CDMA. GLONASS also works on a different spheroid model but its difference from the WGS84 is less than 40cm. Using both GLONASS and GPS together, GNSS receivers have access to 50 satellites and therefore can provide better position location faster, particularly in urban canyons and rugged mountainous areas.

Galileo – European Union

This project is being established by ESA under the leadership of Germany, France and Italy. Other European countries who have joined the project are Norway and Switzerland. Israel, Morocco and Ukraine have also joined the program. China was to have joined but later withdrew as it decided to proceed with its own BeiDou program. The program seeks to develop independent capability in the civilian space as opposed to GPS, BeiDou and GLONASS which are under military control. Galileo will have accuracy higher than GPS and GLONASS at one meter for public use and one centimeter for paid users. There is a provision for exclusive military use in case of extreme situations like war. The program was beset by funding problems as well as friction with US on security issues. However, these have been overcome and 12 of the planned 30 satellites are in orbit. Full operational status is expected by 2020.

Galileo is unique as it is designed to provide a new global search and rescue (SAR) function as part of the MEOSAR system. Apart from relaying the distress signals from emergency beacons to the rescue coordination centre, it will also provide a return link to the beacon to confirm receipt of the distress signal and launch of rescue efforts. This does not exist in the current Cospas Sarsat system.

System	GPS	GLONASS	BeiDou	Galileo	NAVIC
Owner	United States	Russian Federation	China	European Union	India
Orbital altitude	20, 180 km (12, 540 mi)	19, 130 km, (11, 890 mi)	21, 150 km, (13, 140 mi)	23,222 km , (14, 429 mi)	36, 000 km, (22,000mi)
Period	11.97 h, (11 h 58 min)	11.26 h, (11 h 16 min)	12.63 h, (12 h 38 min)	14.08 h, (14h 5min)	
Number of satellites	32 (at least 24 by design)	28 (at least 24 by design) including: 24 operational 2 under check by the satellite prime contractor 2 in flight tests phase	5 geostationary orbit (GEO) satellites, 30 medium Earth orbit (MEO) satellites	4 in-orbit validation satellites + 8 full operation capable satellites in orbit 22 operational satellites budgeted	Total: 7 In Orbit: 7
Frequency	1.57542 GHz (L1 signal) 1.2276 GHz (L2 signal)	Around 1.602 GHz (SP) Around 1.246 GHz (SP)	1.561098 GHz (B1) 1.589742GHz (B1-2) 1.20714 GHz (B2) 1.26852 GHz (B3)	1.164-1.215 GHz (E5a and E5b) 1.260-1.300 GHz (E6) 1.559-1.592 GHz (E2-L1-E11)	L5-band 1164.45-1188.45 MHz S-band 2483.5-2500 MHz
Status	Operational	Operational	22 satellites operational, 40 additional satellites 2016-2010	8 satellites operational, 22 additional satellites 2016-2020	Operational

BDS / BeiDou-2 / COMPASS – China

The Chinese system began as a regional system, BeiDou-1 with three geostationary satellites serving China and its neighborhood. The signals were in the S-band and offered an accuracy of 20m. BeiDou-1 was decommissioned by end of 2012.

BeiDou-2 is a completely new system consisting of 35 satellites of which five are planned to be in geostationary orbit to provide BeiDou-1 compatibility to existing ground systems, 25 are planned in MEO and 5 are planned in inclined geosynchronous orbits. From records of launches there are 6 in geostationary orbit, 8 in inclined geosynchronous orbit and 5 in MEO. The extra satellites in GEO and IGSO are possibly in orbit spares. The BeiDou-2 frequencies overlap the Galileo frequencies, thus providing interoperability between the two systems. As of now, the system is referred as the BeiDou Navigation System, BDS, and currently serves China and its neighborhood. BeiDou is expected to become operational globally by 2020 providing 10m accuracy for public and higher for military use.

Regional Systems

NAVIC – India

The IRNSS system, christened NAVIC after operationalization, is a direct outcome of the restrictions placed on use of the GPS system by the US during the Kargil War with Pakistan. The NAVIC system consists of 7 satellites, 3 in geostationary orbit and 4 in inclined geosynchronous orbits. The system became operational in 2016. The system works in the L5 and S bands and provides better than 20m over the Indian Ocean region and 10m accuracy over India for public use and 10cm for authorized users, which includes the military.

QZSS – Japan

The QZSS or the Quasi-Zenith Satellite System has a unique configuration of the satellite orbit. It is a highly inclined slightly elliptical geosynchronous orbit which gives coverage over Japan and the Pacific region. As of now one satellite, MICHIBIKI has been launched, three more QZSS and one geostationary satellites are in the pipeline. Using the standard GNSS L1, L2 and L5 frequencies, QZSS will complement GPS and will provide positioning, augmented positioning and messaging services. Service is expected to begin from 2018. Ultimately, four more QZSS satellites are planned by 2020. Accuracy is expected to be better than 7m. The precision service will be around 2m and will use a DGPS signal on the L band.

Satellite Based Augmented Systems – SBAS

Based on these GNSS, there are several regional augmented navigation systems under the generic name Satellite Based Augmented System, SBAS. These are Wide Area Augmented System, WAAS in the US, European Geostationary Navigation Overlay Service, EGNOS in Europe and GPS-Aided GEO Augmented Navigation System, GAGAN in India, all based on the GPS system. MTSAT Satellite Based Augmentation Navigation System, MSAS, in Japan and System for Differential Corrections and Monitoring, SDCM, in Russia based on GLONASS and GPS are others. China is planning SNAS, Satellite Navigation Augmentation System, to provide WAAS-like service for the China region. These systems are primarily meant for aircraft navigation including landing and takeoff stages but are also used for other land and sea based navigation applications. They require additional equipment to augment the basic GNSS receivers.

The SBAS is an improvement of the positioning service of the GNSS. The problem with the GNSS is the error due to the atmosphere through which the signal has to pass before reaching the user equipment like a smartphone. This error, called the tropospheric and ionospheric error introduces random variations in the signal which degrades the accuracy in the measurement of the distance of the device from the satellite emitting the signal. Using a GNSS in a differential mode like DGPS can reduce this error. But, this is not possible in a fast moving object like an aircraft. In a SBAS, there are several ground stations which continuously measure the error and update corrections through another channel, like a geostationary satellite. These corrections are applied in real time and can bring down the error to one meter allowing for a CAT-I Instrument Landing System capability.

Other Applications

Though GNSS began as a military system in the US and erstwhile USSR mainly for warship navigation and later extended to ICBMs, it soon became clear that there could be many applications in the civilian area. It took the disastrous Korean Airline flight 007 shooting in 1983 to convince the US that the system needed to be made available for civilian use and further, in 2000 the selective availability was turned off.

Though the US and Russia still exercise military control, new systems like Galileo, NAVIC, BDS and QZSS are primarily civil application driven. Applications of GNSS range from navigation and position location to precision agriculture, surveying, precision timing and scientific studies of the atmosphere and the earth's crust including earthquake prediction.

Business Prospects

Issue 4 of the GNSS Market Report by the European GNSS Agency provides a very good picture of the GNSS marketplace. The study includes both core markets like personal navigation devices as well as assisted markets like smartphones, aviation, precision agriculture and search and rescue. The number of GNSS devices is expected to reach 7 billion, almost one per person by 2019. Smartphone lead the market with 3.08 billion followed by road navigation devices at 0.26 billion as of 2014. The expected growth area is the Asia-Pacific at 11% per annum such that by 2023 it will overtake the combined US and European markets. Middle East and Africa will show a growth rate of 19% as it starts from a low base.

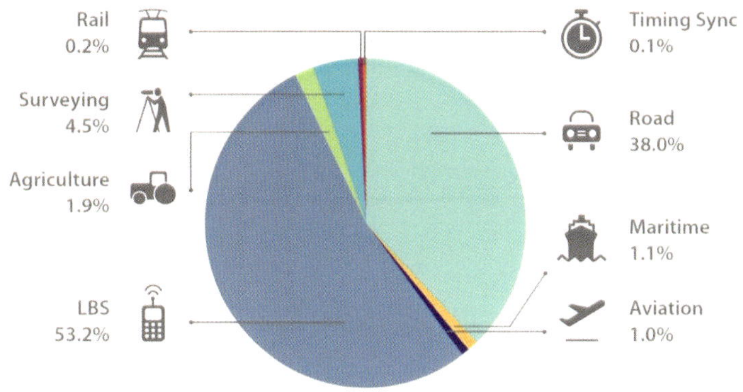

Cumulative core revenue 2013-2023

In smart cities, GNSS will enable smart mobility and LBS applications as well as autonomous driving solutions, travel optimization and automatic transactions such as toll and parking charges. GNSS will be used in container and other mobile asset management. Big Data Analytics will use GNSS data for traffic modeling and management as well as crowd control. With the increased machine to machine interactions GNSS will become an invaluable part of the Internet of Things.

The core market (chip sets) will grow at 8.3% but will slow down to 4.6% from 2020 to 2023 due to lowering of costs but the downstream market of GNSS enabled devices will grow at 7%, beating the Global GDP forecast of 6.6%. LBS will hold the lion's share of 53.2%, followed by road navigation of 38%. Other major markets will be surveying (4.5%) and agriculture at 1.9%. Maritime, aviation, rail and timing will contribute the rest. While GPS is the lead GNSS followed by GLONASS, the focus is on how many devices support multiple GNSS constellations. While devices which support only one GNSS constellation are about 40%, the rest provide at least two GNSS capability with 21% supporting all four global systems, GPS, GLONASS, Galileo and BeiDou. This is an indicator of the future course to be followed by chip makers and downstream device manufacturers and value adders.

The Road Ahead

GNSS have proliferated into four major global and two regional systems. The military origin and the need to have independent military capability have driven this growth. With the opening of the system to civilian use there has been an explosion of applications which have democratized the system. New applications are being generated at a fast pace and this is a major growth area, particularly for the Asia-Pacific region.

Global Positioning System

Global Positioning System or GPS is a satellite navigation system used to determine the ground position of an object. GPS technology was first used by the United States military in the 1960s and expanded into civilian use over the next few decades. Today, GPS receivers are included in many commercial products, such as automobiles, smartphones, exercise watches, and GIS devices.

The GPS system includes 24 satellites deployed in space about 12,000 miles (19,300 kilometers) above the earth's surface. They orbit the earth once every 12 hours at an extremely fast pace of roughly 7,000 miles per hour (11,200 kilometers per hour). The satellites are evenly spread out so that four satellites are accessible via direct line-of-sight from anywhere on the globe.

Each GPS satellite broadcasts a message that includes the satellite's current position, orbit, and exact time. A GPS receiver combines the broadcasts from multiple satellites to calculate its exact position using a process called triangulation. Three satellites are required in order to determine a receiver's location, though a connection to four satellites is ideal since it provides greater accuracy.

In order for a GPS device to work correctly, it must first establish a connection to the required number of satellites. This process can take anywhere from a few seconds to a few minutes, depending on the strength of the receiver. For example, a car's GPS unit will typically establish a GPS connection faster than the receiver in a watch or smartphone. Most GPS devices also use some type of location caching to speed up GPS detection. By memorizing its previous location, a GPS device can quickly determine what satellites will be available the next time it scans for a GPS signal.

Basic Concept of GPS

The GPS concept is based on time and the known position of GPS specialized satellites. The satellites carry very stable atomic clocks that are synchronized with one another and with the ground clocks. Any drift from true time maintained on the ground is corrected daily. In the same manner, the satellite locations are known with great precision. GPS receivers have clocks as well, but they are less stable and less precise.

GPS satellites continuously transmit data about their current time and position. A GPS receiver monitors multiple satellites and solves equations to determine the precise position of the receiver and its deviation from true time. At a minimum, four satellites must be in view of the receiver for it to compute four unknown quantities (three position coordinates and clock deviation from satellite time).

More Detailed Description

Each GPS satellite continually broadcasts a signal (carrier wave with modulation) that includes:

- A pseudorandom code (sequence of ones and zeros) that is known to the receiver. By time-aligning a receiver-generated version and the receiver-measured version of the code, the time of arrival (TOA) of a defined point in the code sequence, called an epoch, can be found in the receiver clock time scale
- A message that includes the time of transmission (TOT) of the code epoch (in GPS time scale) and the satellite position at that time
- Conceptually, the receiver measures the TOAs (according to its own clock) of four satellite signals. From the TOAs and the TOTs, the receiver forms four time of flight (TOF) values, which are (given the speed of light) approximately equivalent to receiver-satellite ranges. The receiver then computes its three-dimensional position and clock deviation from the four TOFs.
- In practice the receiver position (in three dimensional Cartesian coordinates with origin at the Earth's center) and the offset of the receiver clock relative to the GPS time are computed simultaneously, using the navigation equations to process the TOFs.

The receiver's Earth-centered solution location is usually converted to latitude, longitude and height relative to an ellipsoidal Earth model. The height may then be further converted to height relative to the geoid (e.g., EGM96) (essentially, mean sea level). These coordinates may be displayed, e.g., on a moving map display, and/or recorded and/or used by some other system (e.g., a vehicle guidance system).

User-satellite Geometry

Although usually not formed explicitly in the receiver processing, the conceptual time differences of arrival (TDOAs) define the measurement geometry. Each TDOA corresponds to a hyperboloid of revolution. The line connecting the two satellites involved (and its extensions) forms the axis of the hyperboloid. The receiver is located at the point where three hyperboloids intersect.

It is sometimes incorrectly said that the user location is at the intersection of three spheres. While simpler to visualize, this is only the case if the receiver has a clock synchronized with the satellite clocks (i.e., the receiver measures true ranges to the satellites rather than range differences). There are significant performance benefits to the user carrying a clock synchronized with the satellites. Foremost is that only three satellites are needed to compute a position solution. If this were an essential part of the GPS concept so that all users needed to carry a synchronized clock, then a smaller number of satellites could be deployed. However, the cost and complexity of the user equipment would increase significantly.

Receiver in Continuous Operation

The description above is representative of a receiver start-up situation. Most receivers have a track algorithm, sometimes called a *tracker*, that combines sets of satellite measurements collected at different times—in effect, taking advantage of the fact that successive receiver positions are usually close to each other. After a set of measurements are processed, the tracker predicts the receiver location corresponding to the next set of satellite measurements. When the new measurements are collected, the receiver uses a weighting scheme to combine the new measurements with the tracker prediction. In general, a tracker can (a) improve receiver position and time accuracy, (b) reject bad measurements, and (c) estimate receiver speed and direction.

The disadvantage of a tracker is that changes in speed or direction can only be computed with a delay, and that derived direction becomes inaccurate when the distance traveled between two position measurements drops below or near the random error of position measurement. GPS units can use measurements of the Doppler shift of the signals received to compute velocity accurately. More advanced navigation systems use additional sensors like a compass or an inertial navigation system to complement GPS.

Non-navigation Applications

In typical GPS operation as a navigator, four or more satellites must be visible to obtain an accurate result. The solution of the navigation equations gives the position of the receiver along with the difference between the time kept by the receiver's on-board clock and the true time-of-day, thereby eliminating the need for a more precise and possibly impractical receiver based clock. Applications for GPS such as time transfer, traffic signal timing, and synchronization of cell phone base stations, make use of this cheap and highly accurate timing. Some GPS applications use this time for display, or, other than for the basic position calculations, do not use it at all.

Although four satellites are required for normal operation, fewer apply in special cases. If one variable is already known, a receiver can determine its position using only three satellites. For example, a ship or aircraft may have known elevation. Some GPS receivers may use additional clues or assumptions such as reusing the last known altitude, dead reckoning, inertial navigation, or including information from the vehicle computer, to give a (possibly degraded) position when fewer than four satellites are visible.

Structure

The current GPS consists of three major segments. These are the space segment, a control segment, and a user segment. The U.S. Air Force develops, maintains, and operates the space and control segments. GPS satellites broadcast signals from space, and each GPS receiver uses these signals to calculate its three-dimensional location (latitude, longitude, and altitude) and the current time.

Space Segment

Unlaunched GPS block II-A satellite on display at the San Diego Air & Space Museum

The space segment (SS) is composed of 24 to 32 satellites in medium Earth orbit and also includes the payload adapters to the boosters required to launch them into orbit.

The space segment (SS) is composed of the orbiting GPS satellites, or Space Vehicles (SV) in GPS parlance. The GPS design originally called for 24 SVs, eight each in three approximately circular orbits, but this was modified to six orbital planes with four satellites each. The six orbit planes have approximately 55° inclination (tilt relative to the Earth's equator) and are separated by 60° right ascension of the ascending node (angle along the equator from a reference point to the orbit's intersection). The orbital period is one-half a sidereal day, i.e., 11 hours and 58 minutes so that the satellites pass over the same locations or almost the same locations every day. The orbits are arranged so that at least six satellites are always within line of sight from almost everywhere on the Earth's surface. The result of this objective is that the four satellites are not evenly spaced (90°) apart within each orbit. In general terms, the angular difference between satellites in each orbit is 30°, 105°, 120°, and 105° apart, which sum to 360°.

Orbiting at an altitude of approximately 20,200 km (12,600 mi); orbital radius of approximately 26,600 km (16,500 mi), each SV makes two complete orbits each sidereal day, repeating the same ground track each day. This was very helpful during development because even with only four satellites, correct alignment means all four are visible from one spot for a few hours each day. For military operations, the ground track repeat can be used to ensure good coverage in combat zones.

As of February 2016, there are 32 satellites in the GPS constellation, 31 of which are in use. The additional satellites improve the precision of GPS receiver calculations by providing redundant measurements. With the increased number of satellites, the constellation was changed to a non-uniform arrangement. Such an arrangement was shown to improve reliability and availability of the system, relative to a uniform system, when multiple satellites fail. About nine satellites are visible from any point on the ground at any one time, ensuring considerable redundancy over the minimum four satellites needed for a position.

Control Segment

Ground monitor station

The control segment (CS) is composed of:

1. a master control station (MCS),
2. an alternative master control station,
3. four dedicated ground antennas, and
4. six dedicated monitor stations.

The MCS can also access U.S. Air Force Satellite Control Network (AFSCN) ground antennas (for additional command and control capability) and NGA (National Geospatial-Intelligence Agency) monitor stations. The flight paths of the satellites are tracked by dedicated U.S. Air Force monitoring stations in Hawaii, Kwajalein Atoll, Ascension Island, Diego Garcia, Colorado Springs, Colorado and Cape Canaveral, along with shared NGA monitor stations operated in England, Argentina, Ecuador, Bahrain, Australia and Washington DC. The tracking information is sent to the Air Force Space Command MCS at Schriever Air Force Base 25 km (16 mi) ESE of Colorado Springs, which is operated by the 2nd Space Operations Squadron (2 SOPS) of the U.S. Air Force. Then 2 SOPS contacts each GPS satellite regularly with a navigational update using dedicated or shared (AFSCN) ground antennas (GPS dedicated ground antennas are located at Kwajalein, Ascension Island, Diego Garcia, and Cape Canaveral). These updates synchronize the atomic clocks on board the satellites to within a few nanoseconds of each other, and adjust the ephemeris of each satellite's internal orbital model. The updates are created by a Kalman filter that uses inputs from the ground monitoring stations, space weather information, and various other inputs.

Satellite maneuvers are not precise by GPS standards—so to change a satellite's orbit, the satellite must be marked *unhealthy*, so receivers don't use it. After the satellite maneuver, engineers track the new orbit from the ground, upload the new ephemeris, and mark the satellite healthy again.

The operation control segment (OCS) currently serves as the control segment of record. It provides the operational capability that supports GPS users and keeps the GPS operational and performing within specification.

OCS successfully replaced the legacy 1970s-era mainframe computer at Schriever Air Force Base in September 2007. After installation, the system helped enable upgrades and provide a foundation for a new security architecture that supported U.S. armed forces.

OCS will continue to be the ground control system of record until the new segment, Next Generation GPS Operation Control System (OCX), is fully developed and functional. The new capabilities provided by OCX will be the cornerstone for revolutionizing GPS's mission capabilities, enabling Air Force Space Command to greatly enhance GPS operational services to U.S. combat forces, civil partners and myriad domestic and international users. The GPS OCX program also will reduce cost, schedule and technical risk. It is designed to provide 50% sustainment cost savings through efficient software architecture and Performance-Based Logistics. In addition, GPS OCX is expected to cost millions less than the cost to upgrade OCS while providing four times the capability.

The GPS OCX program represents a critical part of GPS modernization and provides significant information assurance improvements over the current GPS OCS program.

- OCX will have the ability to control and manage GPS legacy satellites as well as the next generation of GPS III satellites, while enabling the full array of military signals.
- Built on a flexible architecture that can rapidly adapt to the changing needs of today's and future GPS users allowing immediate access to GPS data and constellation status through secure, accurate and reliable information.
- Provides the warfighter with more secure, actionable and predictive information to enhance situational awareness.
- Enables new modernized signals (L1C, L2C, and L5) and has M-code capability, which the legacy system is unable to do.
- Provides significant information assurance improvements over the current program including detecting and preventing cyber attacks, while isolating, containing and operating during such attacks.
- Supports higher volume near real-time command and control capabilities and abilities.
- On September 14, 2011, the U.S. Air Force announced the completion of GPS OCX Preliminary Design Review and confirmed that the OCX program is ready for the next phase of development.

The GPS OCX program has missed major milestones and is pushing the GPS IIIA launch beyond April 2016.

User Segment

GPS receivers come in a variety of formats, from devices integrated into cars, phones, and watches, to dedicated devices such as these.

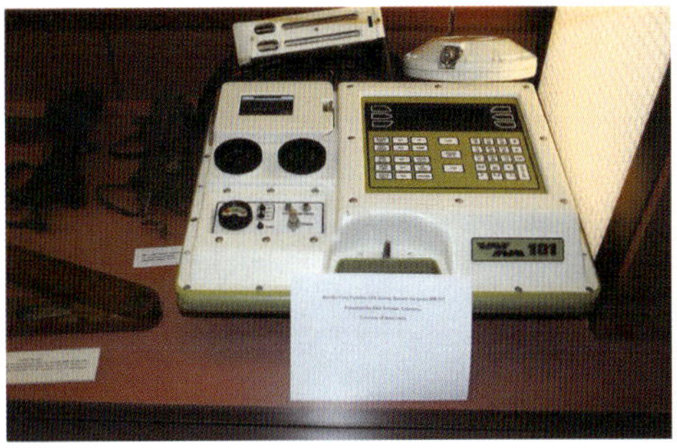

The first portable GPS unit, Leica WM 101 displayed at the Irish National Science Museum at Maynooth.

The user segment (US) is composed of hundreds of thousands of U.S. and allied military users of the secure GPS Precise Positioning Service, and tens of millions of civil, commercial and scientific users of the Standard Positioning Service. In general, GPS receivers are composed of an antenna, tuned to the frequencies transmitted by the satellites, receiver-processors, and a highly stable clock (often a crystal oscillator). They may also include a display for providing location and speed information to the user. A receiver is often described by its number of channels: this signifies how many satellites it can monitor simultaneously. Originally limited to four or five, this has progressively increased over the years so that, as of 2007, receivers typically have between 12 and 20 channels. Though there are many receiver manufacturers, they almost all use one of the chipsets produced for this purpose.

A typical OEM GPS receiver module measuring 15×17 mm.

GPS receivers may include an input for differential corrections, using the RTCM SC-104 format. This is typically in the form of an RS-232 port at 4,800 bit/s speed. Data is actually sent at a much lower rate, which limits the accuracy of the signal sent using RTCM. Receivers with internal DGPS receivers can outperform those using external RTCM data. As of 2006, even low-cost units commonly include Wide Area Augmentation System (WAAS) receivers.

Many GPS receivers can relay position data to a PC or other device using the NMEA 0183 protocol. Although this protocol is officially defined by the National Marine Electronics Association (NMEA), references to this protocol have been compiled from public records, allowing open source tools like gpsd to read the protocol without violating intellectual property laws. Other proprietary protocols exist as well, such as the SiRFand MTK protocols. Receivers can interface with other devices using methods including a serial connection, USB, or Bluetooth.

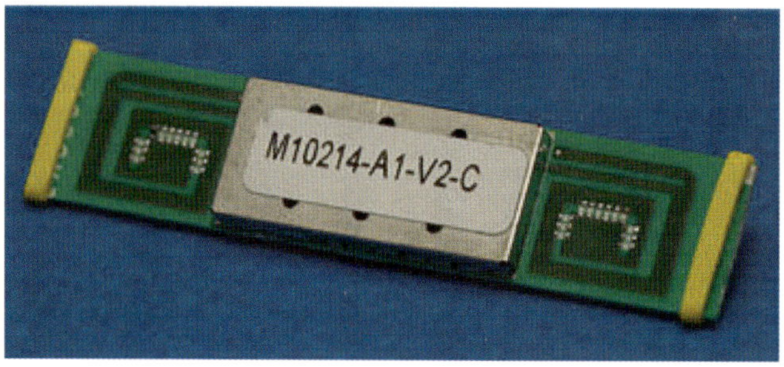

A typical GPS receiver with integrated antenna.

Applications

While originally a military project, GPS is considered a *dual-use* technology, meaning it has significant military and civilian applications.

GPS has become a widely deployed and useful tool for commerce, scientific uses, tracking, and surveillance. GPS's accurate time facilitates everyday activities such as banking, mobile phone operations, and even the control of power grids by allowing well synchronized hand-off switching.

Civilian

This antenna is mounted on the roof of a hut containing a scientific experiment needing precise timing.

Many civilian applications use one or more of GPS's three basic components: absolute location, relative movement, and time transfer.

- Agriculture: GPS has made a great evolution in different aspects of modern agricultural sectors. Today, a growing number of crop producers are using GPS and other modern electronic and computer equipment to practice Site Specific Management (SSM) and precision agriculture. This technology has the potential in agricultural mechanization (farm and machinery management) by providing farmers with a sophisticated tool to measure yield on much smaller scales as well as precise determination and automatic storing of variables such as field time, working area, machine travel distance and speed, fuel consumption and yield information.

- Astronomy: both positional and clock synchronization data is used in astrometry and celestial mechanics. GPS is also used in both amateur astronomy with small telescopes well as by professional observatories for finding extrasolar planets.
- Automated vehicle: applying location and routes for cars and trucks to function without a human driver.
- Cartography: both civilian and military cartographers use GPS extensively.
- Cellular telephony: clock synchronization enables time transfer, which is critical for synchronizing its spreading codes with other base stations to facilitate inter-cell handoff and support hybrid GPS/cellular position detection for mobile emergency calls and other applications. The first handsets with integrated GPS launched in the late 1990s. The U.S. Federal Communications Commission (FCC) mandated the feature in either the handset or in the towers (for use in triangulation) in 2002 so emergency services could locate 911 callers. Third-party software developers later gained access to GPS APIs from Nextel upon launch, followed by Sprint in 2006, and Verizon soon thereafter.
- Clock synchronization: the accuracy of GPS time signals (±10 ns) is second only to the atomic clocks they are based on, and is used in applications such as GPS disciplined oscillators.
- Disaster relief/emergency services: many emergency services depend upon GPS for location and timing capabilities.
- GPS-equipped radiosondes and dropsondes: measure and calculate the atmospheric pressure, wind speed and direction up to 27 km from the Earth's surface.
- Radio occultation for weather and atmospheric science applications.
- Fleet tracking: used to identify, locate and maintain contact reports with one or more fleet vehicles in real-time.
- Geofencing: vehicle tracking systems, person tracking systems, and pet tracking systems use GPS to locate devices that are attached to or carried by a person, vehicle, or pet. The application can provide continuous tracking and send notifications if the target leaves a designated (or "fenced-in") area.
- Geotagging: applies location coordinates to digital objects such as photographs (in Exif data) and other documents for purposes such as creating map overlays with devices like Nikon GP-1
- GPS aircraft tracking
- GPS for mining: the use of RTK GPS has significantly improved several mining operations such as drilling, shoveling, vehicle tracking, and surveying. RTK GPS provides centimeter-level positioning accuracy.
- GPS data mining: It is possible to aggregate GPS data from multiple users to understand movement patterns, common trajectories and interesting locations.

- GPS tours: location determines what content to display; for instance, information about an approaching point of interest.
- Navigation: navigators value digitally precise velocity and orientation measurements.
- Phasor measurements: GPS enables highly accurate timestamping of power system measurements, making it possible to compute phasors.
- Recreation: for example, Geocaching, Geodashing, GPS drawing, waymarking, and other kinds of location based mobile games.
- Robotics: self-navigating, autonomous robots using a GPS sensors, which calculate latitude, longitude, time, speed, and heading.
- Sport: used in football and rugby for the control and analysis of the training load.
- Surveying: surveyors use absolute locations to make maps and determine property boundaries.
- Tectonics: GPS enables direct fault motion measurement of earthquakes. Between earthquakes GPS can be used to measure crustal motion and deformation to estimate seismic strain buildup for creating seismic hazard maps.
- Telematics: GPS technology integrated with computers and mobile communications technology in automotive navigation systems.

Restrictions on Civilian use

The U.S. government controls the export of some civilian receivers. All GPS receivers capable of functioning above 18 km (60,000 feet) altitude and 515 m/s (1,000 knots), or designed or modified for use with unmanned air vehicles like, e.g., ballistic or cruise missile systems, are classified as munitions (weapons)—which means they require State Department export licenses.

This rule applies even to otherwise purely civilian units that only receive the L1 frequency and the C/A (Coarse/Acquisition) code.

Disabling operation above these limits exempts the receiver from classification as munition. Vendor interpretations differ. The rule refers to operation at both the target altitude and speed, but some receivers stop operating even when stationary. This has caused problems with some amateur radio balloon launches that regularly reach 30 km (100,000 feet).

These limits only apply to units or components exported from the United States. A growing trade in various components exists, including GPS units from other countries. These are expressly sold as ITAR-free.

Military

As of 2009, military GPS applications include:

- Navigation: Soldiers use GPS to find objectives, even in the dark or in unfamiliar territory, and to coordinate troop and supply movement. In the United States armed forces, commanders use the *Commander's Digital Assistant* and lower ranks use the Soldier *Digital Assistant*.

Attaching a GPS guidance kit to a dumb bomb, March 2003.

M982 Excalibur GPS-guided artillery shell.

- Target tracking: Various military weapons systems use GPS to track potential ground and air targets before flagging them as hostile. These weapon systems pass target coordinates to precision-guided munitions to allow them to engage targets accurately. Military aircraft, particularly in air-to-groundroles, use GPS to find targets.
- Missile and projectile guidance: GPS allows accurate targeting of various military weapons including ICBMs, cruise missiles, precision-guided munitions and artillery shells. Embedded GPS receivers able to withstand accelerations of 12,000 *g* or about 118 km/s^{2}have been developed for use in 155-millimeter (6.1 in) howitzer shells.
- Search and rescue.
- Reconnaissance: Patrol movement can be managed more closely.
- GPS satellites carry a set of nuclear detonation detectors consisting of an optical sensor called a bhangmeter, an X-ray sensor, a dosimeter, and an electromagnetic pulse (EMP) sensor (W-sensor), that form a major portion of the United States Nuclear Detonation Detection System. General William Shelton has stated that future satellites may drop this feature to save money.

GPS type navigation was first used in war in the 1991 Persian Gulf War, before GPS was fully developed in 1995, to assist Coalition Forces to navigate and perform maneuvers in the war. The war also demonstrated the vulnerability of GPS to being jammed, when Iraqi forces installed jamming devices on likely targets that emitted radio noise, disrupting reception of the weak GPS signal.

Communication

The navigational signals transmitted by GPS satellites encode a variety of information including satellite positions, the state of the internal clocks, and the health of the network. These signals are transmitted on two separate carrier frequencies that are common to all satellites in the network. Two different encodings are used: a public encoding that enables lower resolution navigation, and an encrypted encoding used by the U.S. military.

Message Format

GPS message format	
Subframes	Description
1	Satellite clock, GPS time relationship

2–3	Ephemeris (precise satellite orbit)
4–5	Almanac component (satellite network synopsis, error correction)

Each GPS satellite continuously broadcasts a *navigation message* on L1 (C/A and P/Y) and L2 (P/Y) frequencies at a rate of 50 bits per second. Each complete message takes 750 seconds (12 1/2 minutes) to complete. The message structure has a basic format of a 1500-bit-long frame made up of five subframes, each subframe being 300 bits (6 seconds) long. Subframes 4 and 5 are subcommutated 25 times each, so that a complete data message requires the transmission of 25 full frames. Each subframe consists of ten words, each 30 bits long. Thus, with 300 bits in a subframe times 5 subframes in a frame times 25 frames in a message, each message is 37,500 bits long. At a transmission rate of 50-bit/s, this gives 750 seconds to transmit an entire almanac message (GPS). Each 30-second frame begins precisely on the minute or half-minute as indicated by the atomic clock on each satellite.

The first subframe of each frame encodes the week number and the time within the week, as well as the data about the health of the satellite. The second and the third subframes contain the *ephemeris* – the precise orbit for the satellite. The fourth and fifth subframes contain the *almanac*, which contains coarse orbit and status information for up to 32 satellites in the constellation as well as data related to error correction. Thus, to obtain an accurate satellite location from this transmitted message, the receiver must demodulate the message from each satellite it includes in its solution for 18 to 30 seconds. To collect all transmitted almanacs, the receiver must demodulate the message for 732 to 750 seconds or 12 1/2 minutes.

All satellites broadcast at the same frequencies, encoding signals using unique code division multiple access (CDMA) so receivers can distinguish individual satellites from each other. The system uses two distinct CDMA encoding types: the coarse/acquisition (C/A) code, which is accessible by the general public, and the precise (P(Y)) code, which is encrypted so that only the U.S. military and other NATO nations who have been given access to the encryption code can access it.

The ephemeris is updated every 2 hours and is generally valid for 4 hours, with provisions for updates every 6 hours or longer in non-nominal conditions. The almanac is updated typically every 24 hours. Additionally, data for a few weeks following is uploaded in case of transmission updates that delay data upload.

Satellite Frequencies

GPS frequency overview:607		
Band	Frequency	Description
L1	1575.42 MHz	Coarse-acquisition (C/A) and encrypted precision (P(Y)) codes, plus the L1 civilian (L1C) and military (M) codes on future Block III satellites.
L2	1227.60 MHz	P(Y) code, plus the L2C and military codes on the Block IIR-M and newer satellites.
L3	1381.05 MHz	Used for nuclear detonation (NUDET) detection.
L4	1379.913 MHz	Being studied for additional ionospheric correction.
L5	1176.45 MHz	Proposed for use as a civilian safety-of-life (SoL) signal.

All satellites broadcast at the same two frequencies, 1.57542 GHz (L1 signal) and 1.2276 GHz (L2 signal). The satellite network uses a CDMA spread-spectrum technique where the low-bitrate message data is encoded with a high-rate pseudo-random (PRN) sequence that is different for each satellite. The receiver must be aware of the PRN codes for each satellite to reconstruct the actual message data. The C/A code, for civilian use, transmits data at 1.023 million chips per second, whereas the P code, for U.S. military use, transmits at 10.23 million chips per second. The actual internal reference of the satellites is 10.22999999543 MHz to compensate for relativistic effects that make observers on the Earth perceive a different time reference with respect to the transmitters in orbit. The L1 carrier is modulated by both the C/A and P codes, while the L2 carrier is only modulated by the P code. The P code can be encrypted as a so-called P(Y) code that is only available to military equipment with a proper decryption key. Both the C/A and P(Y) codes impart the precise time-of-day to the user.

The L3 signal at a frequency of 1.38105 GHz is used to transmit data from the satellites to ground stations. This data is used by the United States Nuclear Detonation (NUDET) Detection System (USNDS) to detect, locate, and report nuclear detonations (NUDETs) in the Earth's atmosphere and near space. One usage is the enforcement of nuclear test ban treaties.

The L4 band at 1.379913 GHz is being studied for additional ionospheric correction.

The L5 frequency band at 1.17645 GHz was added in the process of GPS modernization. This frequency falls into an internationally protected range for aeronautical navigation, promising little or no interference under all circumstances. The first Block IIF satellite that provides this signal was launched in May 2010. On February 5th 2016, the 12th and final Block IIF satellite was launched. The L5 consists of two carrier components that are in phase quadrature with each other. Each carrier component is bi-phase shift key (BPSK) modulated by a separate bit train. "L5, the third civil GPS signal, will eventually support safety-of-life applications for aviation and provide improved availability and accuracy."

In 2011, a conditional waiver was granted to LightSquared to operate a terrestrial broadband service near the L1 band. Although LightSquared had applied for a license to operate in the 1525 to 1559 band as early as 2003 and it was put out for public comment, the FCC asked LightSquared to form a study group with the GPS community to test GPS receivers and identify issue that might arise due to the larger signal power from the LightSquared terrestrial network. The GPS community had not objected to the LightSquared (formerly MSV and SkyTerra) applications until November 2010, when LightSquared applied for a modification to its Ancillary Terrestrial Component (ATC) authorization. This filing (SAT-MOD-20101118-00239) amounted to a request to run several orders of magnitude more power in the same frequency band for terrestrial base stations, essentially repurposing what was supposed to be a "quiet neighborhood" for signals from space as the equivalent of a cellular network. Testing in the first half of 2011 has demonstrated that the impact of the lower 10 MHz of spectrum is minimal to GPS devices (less than 1% of the total GPS devices are affected). The upper 10 MHz intended for use by LightSquared may have some impact on GPS devices. There is some concern that this may seriously degrade the GPS signal for many consumer uses. Aviation Week magazine reports that the latest testing (June 2011) confirms "significant jamming" of GPS by LightSquared's system.

Demodulation and Decoding

Because all of the satellite signals are modulated onto the same L1 carrier frequency, the signals must be separated after demodulation. This is done by assigning each satellite a unique binary sequence known as a Gold code. The signals are decoded after demodulation using addition of the Gold codes corresponding to the satellites monitored by the receiver.

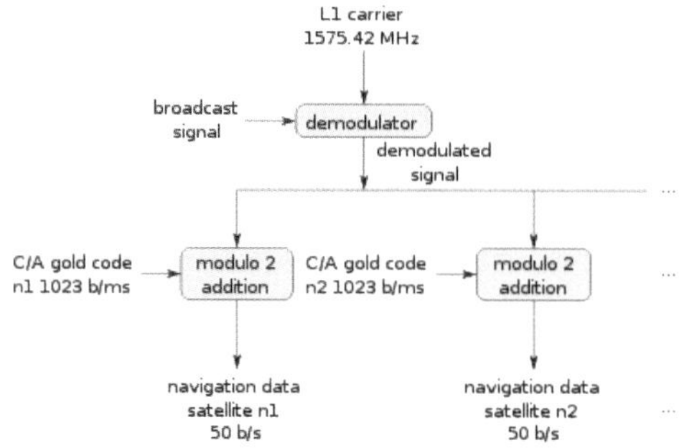

Demodulating and Decoding GPS Satellite Signals using the Coarse/Acquisition Gold code.

If the almanac information has previously been acquired, the receiver picks the satellites to listen for by their PRNs, unique numbers in the range 1 through 32. If the almanac information is not in memory, the receiver enters a search mode until a lock is obtained on one of the satellites. To obtain a lock, it is necessary that there be an unobstructed line of sight from the receiver to the satellite. The receiver can then acquire the almanac and determine the satellites it should listen for. As it detects each satellite's signal, it identifies it by its distinct C/A code pattern. There can be a delay of up to 30 seconds before the first estimate of position because of the need to read the ephemeris data.

Processing of the navigation message enables the determination of the time of transmission and the satellite position at this time.

Navigation Equations

Problem Description

The receiver uses messages received from satellites to determine the satellite positions and time sent. The x, y, and z components of satellite position and the time sent are designated as $[x_i, y_i, z_i, s_i]$ where the subscript i denotes the satellite and has the value 1, 2, ..., n, where $n \geq 4$. When the time of message reception indicated by the on-board receiver clock is t_i, the true reception time is $t_i = t_i - b$, where b is the receiver's clock bias from the much more accurate GPS clocks employed by the satellites. The receiver clock bias is the same for all received satellite signals (assuming the satellite clocks are all perfectly synchronized). The message's transit time is $t_i - b - s_i$, where s_i is the satellite time. Assuming the message traveled at the speed of light, c, the distance traveled is $(t_i - b - s_i)c$.

For n satellites, the equations to satisfy are:

$$d_i = (t_i - b - s_i)c, i = 1,2,\ldots,n$$

where d_i is the geometric distance or range between receiver and satellite:

$$d_i = \sqrt{(x - x_i)^2 + (y - y_i)^2 + (z - z_i)^2}$$

Defining pseudoranges as $p_i = (t_i - s_i)c$, we see they are biased versions of the true range:

$$p_i = d_i + bc, i = 1,2,...,n \ .$$

Since the equations have four unknowns $[x, y, z, b]$—the three components of GPS receiver position and the clock bias—signals from at least four satellites are necessary to attempt solving these equations. They can be solved by algebraic or numerical methods. Existence and uniqueness of GPS solutions are discussed by Abell and Chaffee. When n is greater than 4 this system is overdetermined and a fitting method must be used.

The amount of error in the results varies with the received satellites' locations in the sky, since certain configurations (when the received satellites are close together in the sky) cause large errors. Receivers usually calculate a running estimate of the error in the calculated position. This is done by multiplying the basic resolution of the receiver by quantities called the geometric dilution of position (GDOP) factors, calculated from the relative sky directions of the satellites used. The receiver location is expressed in a specific coordinate system, such as latitude and longitude using the WGS 84 geodetic datum or a country-specific system.

Geometric Interpretation

The GPS equations can be solved by numerical and analytical methods. Geometrical interpretations can enhance the understanding of these solution methods.

Spheres

The measured ranges, called pseudoranges, contain clock errors. In a simplified idealization in which the ranges are synchronized, these true ranges represent the radii of spheres, each centered on one of the transmitting satellites. The solution for the position of the receiver is then at the intersection of the surfaces of three of these spheres. If more than the minimum number of ranges is available, a near intersection of more than three sphere surfaces could be found via, e.g. least squares.

Hyperboloids

If the pseudorange between the receiver and satellite i and the pseudorange between the receiver and satellite j are subtracted, $p_i - p_j$, the common receiver clock bias (b) cancels out, resulting in a difference of distances $d_j - d_j$. The locus of points having a constant difference in distance to two points (here, two satellites) is a hyperbola on a plane and a hyperboloid of revolution in 3D space. Thus, from four pseudorange measurements, the receiver can be placed at the intersection of the surfaces of three hyperboloids each with foci at a pair of satellites. With additional satellites, the multiple intersections are not necessarily unique, and a best-fitting solution is sought instead.

Inscribed Sphere

The receiver position can be interpreted as the center of a inscribed sphere (insphere) of radius *bc*, given by the receiver clock bias *b* (scaled by the speed of light *c*). The insphere location is such that it touches other spheres. The circumscribing spheres are centered at the GPS satellites, whose radii equal the measured pseudoranges p_i. This configuration is distinct from the one described in section Spheres, in which the spheres' radii were the unbiased or geometric ranges d_i.

Spherical Cones

The solution space $[x, y, z, b]$ can be seen as a four-dimensional geometric space. In that case each of the equations describes a spherical cone, with the cusp located at the satellite, and the base a sphere around the satellite. The receiver is at the intersection of four or more of such cones.

Solution Methods

Least Squares

When more than four satellites are available, the calculation can use the four best, or more than four simultaneously (up to all visible satellites), depending on the number of receiver channels, processing capability, and geometric dilution of precision(GDOP).

Using more than four involves an over-determined system of equations with no unique solution; such a system can be solved by a least-squares or weighted least squares method.

$$\left(\hat{x},\hat{y},\hat{z},\hat{b}\right) = \underset{(x,y,z,b)}{\arg\min} \sum_i \left(\sqrt{(x-x_i)^2+(y-y_i)^2+(z-z_i)^2}+bc-p_i\right)^2$$

Iterative

Both the equations for four satellites, or the least squares equations for more than four, are non-linear and need special solution methods. A common approach is by iteration on a linearized form of the equations, such as the Gauss–Newton algorithm.

The GPS was initially developed assuming use of a numerical least-squares solution method—i.e., before closed-form solutions were found.

Closed-form

One closed-form solution to the above set of equations was developed by S. Bancroft. Its properties are well known; in particular, proponents claim it is superior in low-GDOP situations, compared to iterative least squares methods.

Bancroft's method is algebraic, as opposed to numerical, and can be used for four or more satellites. When four satellites are used, the key steps are inversion of a 4x4 matrix and solution of a single-variable quadratic equation. Bancroft's method provides one or two solutions for the unknown quantities. When there are two (usually the case), only one is a near-Earth sensible solution.

When a receiver uses more than four satellites for a solution, Bancroft uses the generalized inverse (i.e., the pseudoinverse) to find a solution. However, a case has been made that iterative methods (e.g., Gauss–Newton algorithm) for solving over-determined non-linear least squares (NLLS) problems generally provide more accurate solutions.

Leick et al. (2015) states that "Bancroft's (1985) solution is a very early, if not the first, closed-form solution." Other closed-form solutions were published afterwards, although their adoption in practice is unclear.

Error Sources and Analysis

GPS error analysis examines error sources in GPS results and the expected size of those errors. GPS makes corrections for receiver clock errors and other effects, but some residual errors remain uncorrected. Error sources include signal arrival time measurements, numerical calculations, atmospheric effects (ionospheric/tropospheric delays), ephemeris and clock data, multipath signals, and natural and artificial interference. Magnitude of residual errors from these sources depends on geometric dilution of precision. Artificial errors may result from jamming devices and threaten ships and aircraft or from intentional signal degradation through selective availability, which limited accuracy to ≈6–12 m, but has been switched off since May 1, 2000.

Accuracy Enhancement and Surveying

Augmentation

Integrating external information into the calculation process can materially improve accuracy. Such augmentation systems are generally named or described based on how the information arrives. Some systems transmit additional error information (such as clock drift, ephemera, or ionospheric delay), others characterize prior errors, while a third group provides additional navigational or vehicle information.

Examples of augmentation systems include the Wide Area Augmentation System (WAAS), European Geostationary Navigation Overlay Service (EGNOS), Differential GPS (DGPS), inertial navigation systems (INS) and Assisted GPS. The standard accuracy of about 15 meters (49 feet) can be augmented to 3–5 meters (9.8–16.4 ft) with DGPS, and to about 3 meters (9.8 feet) with WAAS.

Precise Monitoring

Accuracy can be improved through precise monitoring and measurement of existing GPS signals in additional or alternative ways.

The largest remaining error is usually the unpredictable delay through the ionosphere. The spacecraft broadcast ionospheric model parameters, but some errors remain. This is one reason GPS spacecraft transmit on at least two frequencies, L1 and L2. Ionospheric delay is a well-defined function of frequency and the total electron content (TEC) along the path, so measuring the arrival time difference between the frequencies determines TEC and thus the precise ionospheric delay at each frequency.

Military receivers can decode the P(Y) code transmitted on both L1 and L2. Without decryption keys, it is still possible to use a codeless technique to compare the P(Y) codes on L1 and L2 to gain much of the same error information. However, this technique is slow, so it is currently available only on specialized surveying equipment. In the future, additional civilian codes are expected to be transmitted on the L2 and L5 frequencies. All users will then be able to perform dual-frequency measurements and directly compute ionospheric delay errors.

A second form of precise monitoring is called Carrier-Phase Enhancement (CPGPS). This corrects the error that arises because the pulse transition of the PRN is not instantaneous, and thus the correlation (satellite–receiver sequence matching) operation is imperfect. CPGPS uses the L1 carrier wave, which has a period of $\frac{1\text{s}}{1575.42\times10^6}=0.63475\text{ns}\approx1\text{ns}$, which is about one-thousandth of the C/A Gold code bit period of $\frac{1\text{s}}{1023\times10^3}=977.5\text{ns}\approx1000\text{ns}$, to act as an additional clock signal and resolve the uncertainty. The phase difference error in the normal GPS amounts to 2–3 meters (7–10 ft) of ambiguity. CPGPS working to within 1% of perfect transition reduces this error to 3 centimeters (1.2 in) of ambiguity. By eliminating this error source, CPGPS coupled with DGPS normally realizes between 20–30 centimeters (8–12 in) of absolute accuracy.

Relative Kinematic Positioning (RKP) is a third alternative for a precise GPS-based positioning system. In this approach, determination of range signal can be resolved to a precision of less than 10 centimeters (4 in). This is done by resolving the number of cycles that the signal is transmitted and received by the receiver by using a combination of differential GPS (DGPS) correction data, transmitting GPS signal phase information and ambiguity resolution techniques via statistical tests—possibly with processing in real-time (real-time kinematic positioning, RTK).

Timekeeping

Leap Seconds

While most clocks derive their time from Coordinated Universal Time (UTC), the atomic clocks on the satellites are set to GPS time. The difference is that GPS time is not corrected to match the rotation of the Earth, so it does not contain leap seconds or other corrections that are periodically added to UTC. GPS time was set to match UTC in 1980, but has since diverged. The lack of corrections means that GPS time remains at a constant offset with International Atomic Time (TAI) (TAI – GPS = 19 seconds). Periodic corrections are performed to the on-board clocks to keep them synchronized with ground clocks.

The GPS navigation message includes the difference between GPS time and UTC. As of January 2017, GPS time is 18 seconds ahead of UTC because of the leap second added to UTC on December 31, 2016. Receivers subtract this offset from GPS time to calculate UTC and specific timezone values. New GPS units may not show the correct UTC time until after receiving the UTC offset message. The GPS-UTC offset field can accommodate 255 leap seconds (eight bits).

Accuracy

GPS time is theoretically accurate to about 14 nanoseconds, due to the clock drift that atomic clocks experience in GPS transmitters. However, most receivers lose accuracy in the interpretation of the signals and are only accurate to 100 nanoseconds.

Format

As opposed to the year, month, and day format of the Gregorian calendar, the GPS date is expressed as a week number and a seconds-into-week number. The week number is transmitted as a ten-bit field in the C/A and P(Y) navigation messages, and so it becomes zero again every 1,024 weeks (19.6 years). GPS week zero started at 00:00:00 UTC (00:00:19 TAI) on January 6, 1980, and the week number became zero again for the first time at 23:59:47 UTC on August 21, 1999 (00:00:19 TAI on August 22, 1999). To determine the current Gregorian date, a GPS receiver must be provided with the approximate date (to within 3,584 days) to correctly translate the GPS date signal. To address this concern the modernized GPS navigation message uses a 13-bit field that only repeats every 8,192 weeks (157 years), thus lasting until the year 2137 (157 years after GPS week zero).

Carrier Phase Tracking (Surveying)

Another method that is used in surveying applications is carrier phase tracking. The period of the carrier frequency multiplied by the speed of light gives the wavelength, which is about 0.19 meters for the L1 carrier. Accuracy within 1% of wavelength in detecting the leading edge reduces this component of pseudorange error to as little as 2 millimeters. This compares to 3 meters for the C/A code and 0.3 meters for the P code.

However, 2 millimeter accuracy requires measuring the total phase—the number of waves multiplied by the wavelength plus the fractional wavelength, which requires specially equipped receivers. This method has many surveying applications. It is accurate enough for real-time tracking of the very slow motions of tectonic plates, typically 0–100 mm (0–4 inches) per year.

Triple differencing followed by numerical root finding, and a mathematical technique called least squares can estimate the position of one receiver given the position of another. First, compute the difference between satellites, then between receivers, and finally between epochs. Other orders of taking differences are equally valid. Detailed discussion of the errors is omitted.

The satellite carrier total phase can be measured with ambiguity as to the number of cycles. Let $\phi(r_i, s_j, t_k)$ denote the phase of the carrier of satellite j measured by receiver i at time t_k. This notation shows the meaning of the subscripts i, j, and k. The receiver (r), satellite (s), and time (t) come in alphabetical order as arguments of ϕ and to balance readability and conciseness, let $\phi_{i,j,k} = \phi(r_i, s_j, t_k)$ be a concise abbreviation. Also we define three functions, : $\Delta^r, \Delta^s, \Delta^t$, which return differences between receivers, satellites, and time points, respectively. Each function has variables with three subscripts as its arguments. These three functions are defined below. If $\alpha_{i,j,k}$ is a function of the three integer arguments, i, j, and k then it is a valid argument for the functions, : $\Delta^r, \Delta^s, \Delta^t$, with the values defined as

$$\Delta^r(\alpha_{i,j,k}) = \alpha_{i+1,j,k} - \alpha_{i,j,k},$$

$$\Delta^s(\alpha_{i,j,k}) = \alpha_{i,j+1,k} - \alpha_{i,j,k}, \text{ and}$$

$$\Delta^t(\alpha_{i,j,k}) = \alpha_{i,j,k+1} - \alpha_{i,j,k} \ .$$

Also if $\alpha_{i,j,k}$ *and* $\beta_{l,m,n}$ are valid arguments for the three functions and a and b are constants then $(a\,\alpha_{i,j,k} + b\,\beta_{l,m,n})$ is a valid argument with values defined as

$$\Delta^r(a\,\alpha_{i,j,k} + b\,\beta_{l,m,n}) = a\,\Delta^r(\alpha_{i,j,k}) + b\,\Delta^r(\beta_{l,m,n}),$$

$$\Delta^s(a\,\alpha_{i,j,k} + b\,\beta_{l,m,n}) = a\,\Delta^s(\alpha_{i,j,k}) + b\,\Delta^s(\beta_{l,m,n}),\text{ and}$$

$$\Delta^t(a\,\alpha_{i,j,k} + b\,\beta_{l,m,n}) = a\,\Delta^t(\alpha_{i,j,k}) + b\,\Delta^t(\beta_{l,m,n})\,.$$

Receiver clock errors can be approximately eliminated by differencing the phases measured from satellite 1 with that from satellite 2 at the same epoch. This difference is designated as $\Delta^s(\phi_{1,1,1}) = \phi_{1,2,1} - \phi_{1,1,1}$

Double differencing computes the difference of receiver 1's satellite difference from that of receiver 2. This approximately eliminates satellite clock errors. This double difference is:

$$\Delta^r(\Delta^s(\phi_{1,1,1})) \;=\; \Delta^r(\phi_{1,2,1} - \phi_{1,1,1}) \;=\; \Delta^r(\phi_{1,2,1}) - \Delta^r(\phi_{1,1,1}) = \;(\phi_{2,2,1} - \phi_{1,2,1}) - (\phi_{2,1,1} - \phi_{1,1,1})$$

Triple differencing subtracts the receiver difference from time 1 from that of time 2. This eliminates the ambiguity associated with the integral number of wavelengths in carrier phase provided this ambiguity does not change with time. Thus the triple difference result eliminates practically all clock bias errors and the integer ambiguity. Atmospheric delay and satellite ephemeris errors have been significantly reduced. This triple difference is:

$$\Delta^t(\Delta^r(\Delta^s(\phi_{1,1,1})))$$

Triple difference results can be used to estimate unknown variables. For example, if the position of receiver 1 is known but the position of receiver 2 unknown, it may be possible to estimate the position of receiver 2 using numerical root finding and least squares. Triple difference results for three independent time pairs may be sufficient to solve for receiver 2's three position components. This may require a numerical procedure. An approximation of receiver 2's position is required to use such a numerical method. This initial value can probably be provided from the navigation message and the intersection of sphere surfaces. Such a reasonable estimate can be key to successful multidimensional root finding. Iterating from three time pairs and a fairly good initial value produces one observed triple difference result for receiver 2's position. Processing additional time pairs can improve accuracy, overdetermining the answer with multiple solutions. Least squares can estimate an overdetermined system. Least squares determines the position of receiver 2 that best fits the observed triple difference results for receiver 2 positions under the criterion of minimizing the sum of the squares.

Advantages and Disadvantages of GPS

Every coin has two sides. If there are bene×ts then it is also evident that along with that there must be some limitations or disadvantages. The same rule applies to the GPS systems also. There are a lot of advantages that we avail out of this technology, but somewhere around we have to make some compromise also owing to the limitation of this technology. GPS system is a latest technological leap taken, which has brought new features to the communication technology and various other sphere of our life. GPS has many advantages when it comes to useful features like tracking cars and mobiles or knowing directions while travelling. It is an apt answer to the security concerns we are facing increasingly in today's age. Cars are stolen and with GPS systems, they can well be traced and returned to the user. Apart from that, GPS can be installed in mobile phones and it again proves to be a huge assistance. Also it is a great advantage installing GPS in school or college buses and Cabs for additional security of children and women respectively. But at the same time there are few limitations too. At times its accuracy can be doubted and also there is a fear of being tracked by other device which may lead to security concerns.

Advantages of Global Positioning Systems

This satellite based navigation system is an important tool for military, civil and commercial users. The greatest advantage of this technology is, that it is provided free of cost. There are a lot of advantages associated with this technology. Vehicle tracking systems helps in reducing the expenses by referencing customers who do not realize the risks of high speed and importance of fuel. By evaluation and identifying these drivers, lot of energy can be saves by reducing the fuel consumption. The low reading on speed meter also ensures fewer accidents. However few other advantages include:

- Determines atmosphere's water content, thereby improving the accuracy of the water forecast.
- In traditional methods of survey require a design to make sure that nothing gets in between the measurement line of vision. With the GPS equipment, you do not require an elaborate design of placement and you can select the optimum measurement points despite the terrain of the land.
- Servicing and marinating vehicles requires a lot of time and money. GPS systems help in saving the end costs when it comes to this.
- It helps improve mapping as well as tracking skills and also acts as security against vehicular thefts and for locating someone who got lost.
- Corporate can also reap the advantages of Global Positioning systems. They can keep a check on the productivity of their employees and company as a whole using the GPS technology. GPS can be very effectively used to monitor employees on an individual basis, like their work timings and break timings and can be a highly useful tool to keep a record of the performance of anyone in the firm.

Disadvantages of Global Positioning Systems

If there are so many advantages then it is evident that there must be few disadvantages as well as limitations also. Disadvantages of GPS are that it completely relies on receiving satellite signals by radio, so can be prone to nuclear weapons EMP, radio interference, and failed satellites. These are almost unheard of in the current operation, and let us hope they are unlikely. A lot of users who use GPS in vehicles are prone to accidents if they focus more on GPS rather than road, over that GPS can often land you in a trouble over blocked roads or under construction ones. Another issue is the accuracy. Some potential applications could use even better accuracy, for example as an aid to the blind, automated vehicles and aircraft. Another issue is that the position can be significantly in error occasionally, especially when the number of satellites is restricted. Moreover there is also a risk of cellular devices being tracked by other cellular one's which may lead to security issues.

References

- Evans, James (1998), The History and Practice of Ancient Astronomy, Oxford: Oxford University Press, pp. 102–103, ISBN 9780199874453
- Heiskanen, W. A. (1962). "Is the Earth a triaxial ellipsoid?". Journal of Geophysical Research. 67 (1): 321–327. Bibcode:1962JGR....67..321H. doi:10.1029/JZ067i001p00321
- Heiskanen, W. A. (1962). "Is the Earth a triaxial ellipsoid?". Journal of Geophysical Research. 67 (1): 321–327. Bibcode:1962JGR....67..321H. doi:10.1029/JZ067i001p00321
- American Society of Civil Engineers (1994-01-01). Glossary of the Mapping Sciences. ASCE Publications. p. 224. ISBN 9780784475706
- Kwok, Geodetic Survey Section Lands Department Hong Kong. "Geodetic Datum Transformation, p.24" (PDF). Geodetic Survey Section Lands Department Hong Kong. Retrieved 4 March 2014
- Dziewonski, A. M.; Anderson, D. L. (1981), "Preliminary reference Earth model" (PDF), Physics of the Earth and Planetary Interiors, 25 (4): 297–356, Bibcode:1981PEPI...25..297D, doi:10.1016/0031-9201(81)90046-7, ISSN 0031-9201
- B. Hofmann-Wellenhof; H. Lichtenegger; J. Collins. GPS - theory and practice. Section 10.2.1. p. 282. ISBN 3-211-82839-7

4

Cartography: Study and Practice of Map Making

The science that is concerned with the practice of making maps is called cartography. It is based on the premise that reality can be represented in a way that effectively communicates spatial information. This chapter analyzes the fundamental concepts of terrain cartography, cartographic principles and design, map projection, etc. to provide an easy understanding of cartography.

Maps make our lives infinitely easier. From finding our way around town to finding a store in a shopping mall, maps simplify the physical world. It's not easy to make a map, however. There are countless measurements to be made, lines to be drawn, and all of it must be done with painstaking accuracy. It's no wonder the practice of mapmaking is a complex one.

Cartography, the art and science of graphically representing a geographical area, usually on a flat surface such as a map or chart. It may involve the superimposition of political, cultural, or other nongeographical divisions onto the representation of a geographical area.

Cartography is a complex, an ever-changing field, but at the center of it is the map-making process. Viewed in the broadest sense, this process includes everything from the gathering, evaluation and processing of source data, through the intellectual and graphical design of the map, to the drawing and reproduction of the final document. As such, it is a unique mixture of science, art and technology and calls for a variety of in-depth knowledge and skills on the part of the cartographer. Sometimes one person directs this entire sequence of cartographic activities, but this occurs only in relatively simple cases. In the creation of a map, it is much more common for the various tasks to be split up and accomplished by several individuals.

Cartography is much more than just map-making, however. It is also an academic discipline in its own right. It has its own professional associations (regional, national and international), journals, conferences, educational programs and its own identity. As a discipline, it embraces not only cartographers who make maps, but also cartographers who teach about maps and cartographers who do research on maps. Once seen as the products of a relatively straightforward practical exercise, maps are now viewed as complex intellectual images offering a rich potential for scientific investigation. Whether the thrust of the research is cognitive, mathematical, historical, perceptual or technological, cartographers are exploiting this potential to the fullest

Cartography is an ancient discipline that dates from the prehistoric depiction of hunting and fishing territories. The Babylonians mapped the world in a flattened, disk-shaped form, but

Claudius Ptolemaeus (Ptolemy) established the basis for subsequent efforts in the 2nd century CE with his eight-volume work Geōgraphikē hyphēgēsis (Guide to Geography) that showed a spherical Earth. Maps produced during the Middle Ages followed Ptolemy's guide, but they used Jerusalem as the central feature and placed East at the top. Those representations are often called T-maps because they show only three continents (Europe, Asia, and Africa), separated by the "T" formed by the Mediterranean Sea and the Nile River. More accurate geographical representation began in the 14th century when portolan (seamen's) charts were compiled for navigation.

The discovery of the New World by Europeans led to the need for new techniques in cartography, particularly for the systematic representation on a flat surface of the features of a curved surface—generally referred to as a projection (e.g., Mercator projection, cylindrical projection, and Lambert conformal projection). During the 17th and 18th centuries there was a vast outpouring of printed maps of ever-increasing accuracy and sophistication. Systematic surveys were undertaken involving triangulation that greatly improved map reliability and precision. Noteworthy among the scientific methods introduced later was the use of the telescope for determining the length of a degree of longitude.

Modern cartography largely involves the use of aerial and, increasingly, satellite photographs as a base for any desired map or chart. The procedures for translating photographic data into maps are governed by the principles of photogrammetry and yield a degree of accuracy previously unattainable. The remarkable improvements in satellite photography since the late 20th century and the general availability on the Internet of satellite images have made possible the creation of Google Earth and other databases that are widely available online. Satellite photography has also been used to create highly detailed maps of features of the Moon and of several planets in our solar system and their satellites. In addition, the use of geographic information systems (GIS) has been indispensible in expanding the scope of cartographic subjects.

Cartography today has two essential characteristics. First of all, it is important. Maps perform a fundamental and indispensable role as one of the underpinnings of civilization. Few activities relating to the earth's surface, whether land use planning, property ownership, weather forecasting, road construction, locational analysis, emergency response, forest management, mineral prospecting, navigation–the list is endless–would be practicable without maps. And never has this role been more vital than it is today. Humanity faces severe problems, many of them environmental in nature, and effective mapping is crucial if solutions are to be found. In conjunction with the great data gathering capabilities and analytical power of remote sensing and geographic information systems (GIS), cartography, in many instances, provides the key to finding solutions.

A second, overriding characteristic of cartography today is its dynamic nature. The cartographic discipline is in the throes of a revolution, brought about in large measure by advances in technology and, in particular, by the impact of the computer. Map-making has always been affected by technological change, but the recent transformation of cartography has been unprecedented in scope compared with previous evolutionary changes. While former pen and ink techniques for map-making are still found in isolated use, today's cartographer is more likely to be found seated at a terminal using the very latest in computer hardware and graphic software. In most cases, without any loss in image quality, maps are generated faster with less cost than before, not to mention with even more enjoyment for the map-maker! In essence, the computer equips the cartographer with unparalleled control over the mapping process.

Map Types

General vs. Thematic Cartography

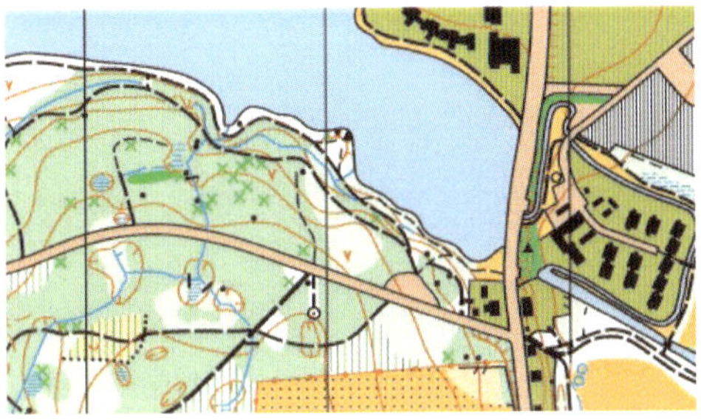
Small section of an orienteering map.

Topographic map of Easter Island.

Relief map Sierra Nevada

In understanding basic maps, the field of cartography can be divided into two general categories: general cartography and thematic cartography. General cartography involves those maps that are constructed for a general audience and thus contain a variety of features. General maps exhibit many reference and location systems and often are produced in a series. For example, the 1:24,000 scale topographic maps of the United States Geological Survey (USGS) are a standard as compared to the 1:50,000 scale Canadian maps. The government of the UK produces the classic 1:50,000 (replacing the older 1 inch to 1 mile) "Ordnance Survey" maps of the entire UK and with a range of correlated larger- and smaller-scale maps of great detail. Many private mapping companies have also produced thematic map series.

Thematic cartography involves maps of specific geographic themes, oriented toward specific audiences. A couple of examples might be a dot map showing corn production in Indiana or a shaded area map of Ohio counties, divided into numerical choropleth classes. As the volume of geographic data has exploded over the last century, thematic cartography has become increasingly useful and necessary to interpret spatial, cultural and social data.

An orienteering map combines both general and thematic cartography, designed for a very specific user community. The most prominent thematic element is shading, that indicates degrees of difficulty of travel due to vegetation. The vegetation itself is not identified, merely classified by the difficulty that it presents.

Topographic vs. Topological

A topographic map is primarily concerned with the topographic description of a place, including (especially in the 20th and 21st centuries) the use of contour lines showing elevation. Terrain or relief can be shown in a variety of ways.

A topological map is a very general type of map, the kind one might sketch on a napkin. It often disregards scale and detail in the interest of clarity of communicating specific route or relational information. Beck's London Underground map is an iconic example. Though the most widely used map of "The Tube," it preserves little of reality: it varies scale constantly and abruptly, it straightens curved tracks, and it contorts directions. The only topography on it is the River Thames, letting the reader know whether a station is north or south of the river. That and the topology of station order and interchanges between train lines are all that is left of the geographic space. Yet those are all a typical passenger wishes to know, so the map fulfills its purpose.

Map Design

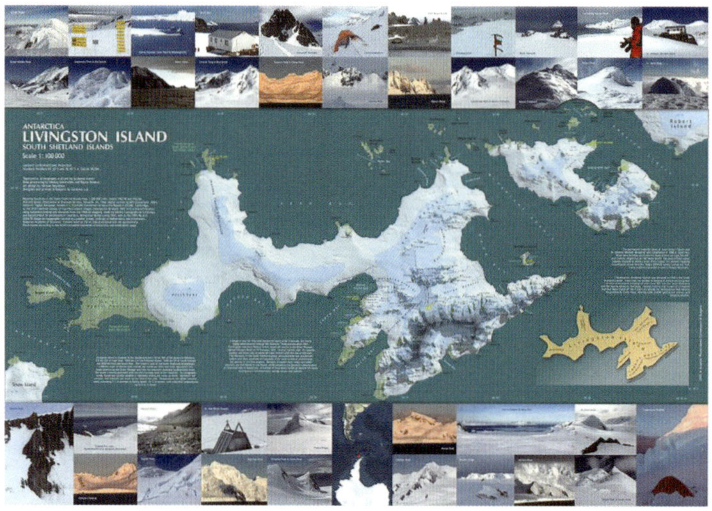

Illustrated map.

Map Purpose and Selection of Information

Arthur H. Robinson, an American cartographer influential in thematic cartography, stated that a map not properly designed "will be a cartographic failure." He also claimed, when considering all aspects of cartography, that "map design is perhaps the most complex." Robinson codified the mapmaker's understanding that a map must be designed foremost with consideration to the audience and its needs.

From the very beginning of mapmaking, maps "have been made for some particular purpose or set of purposes". The intent of the map should be illustrated in a manner in which the percipient acknowledges its purpose in a timely fashion. The term *percipient* refers to the person receiving information and was coined by Robinson. The principle of figure-ground refers to this notion of engaging the user by presenting a clear presentation, leaving no confusion concerning the purpose of the map. This will enhance the user's experience and keep his attention. If the user is unable to identify what is being demonstrated in a reasonable fashion, the map may be regarded as useless.

Making a meaningful map is the ultimate goal. Alan MacEachren explains that a well designed map "is convincing because it implies authenticity" (1994). An interesting map will no doubt engage a reader. Information richness or a map that is multivariate shows relationships within the map. Showing several variables allows comparison, which adds to the meaningfulness of the map. This also generates hypothesis and stimulates ideas and perhaps further research.

In order to convey the message of the map, the creator must design it in a manner which will aid the reader in the overall understanding of its purpose. The title of a map may provide the "needed link" necessary for communicating that message, but the overall design of the map fosters the manner in which the reader interprets it (Monmonier, 1993).

In the 21st century it is possible to find a map of virtually anything from the inner workings of the human body to the virtual worlds of cyberspace. Therefore, there are now a huge variety of different styles and types of map – for example, one area which has evolved a specific and recognisable variation are those used by public transport organisations to guide passengers, namely urban rail and metro maps, many of which are loosely based on 45 degree angles as originally perfected by Harry Beck and George Dow.

Naming Conventions

Most maps use text to label places and for such things as the map title, legend and other information. Although maps are often made in one specific language, place names often differ between languages. So a map made in English may use the name *Germany* for that country, while a German map would use *Deutschland* and a French map *Allemagne*. A non-native term for a place is referred to as an exonym.

In some cases the correct name is not clear. For example, the nation of Burma officially changed its name to Myanmar, but many nations do not recognize the ruling junta and continue to use *Burma*. Sometimes an official name change is resisted in other languages and the older name may remain in common use. Examples include the use of *Saigon* for Ho Chi Minh City, *Bangkok* for Krung Thep and *Ivory Coast* for Côte d'Ivoire.

Difficulties arise when transliteration or transcription between writing systems is required. Some well-known places have well-established names in other languages and writing systems, such as *Russia* or *Rußland* for Росси́я, but in other cases a system of transliteration or transcription is required. Even in the former case, the exclusive use of an exonym may be unhelpful for the map user. It will not be much use for an English user of a map of Italy to show Livorno *only* as "Leghorn" when road signs and railway timetables show it as "Livorno". In transliteration, the characters in one script are represented by characters in another. For example, the Cyrillic letter *P* is usually written as *R* in the Latin script, although in many cases it is not as simple as a one-for-one equivalence. Systems exist for transliteration of Arabic, but the results may vary. For example, the Yemeni city of Mocha is written variously in English as Mocha, Al Mukha, al-Mukhā, Mocca and Moka. Transliteration systems are based on relating written symbols to one another, while transcription is the attempt to spell in one language the phonetic sounds of another. Chinese writing is now usually converted to the Latin alphabet through the Pinyin phonetic transcription systems. Other systems were used in the past, such as Wade-Giles, resulting in the city being spelled *Beijing* on newer English maps and *Peking* on older ones.

Further difficulties arise when countries, especially former colonies, do not have a strong national geographic naming standard. In such cases, cartographers may have to choose between various phonetic spellings of local names versus older imposed, sometimes resented, colonial names. Some countries have multiple official languages, resulting in multiple official placenames. For example, the capital of Belgium is both *Brussel* and *Bruxelles*. In Canada, English and French are official languages and places have names in both languages. British Columbia is also officially named *la Colombie-Britannique*. English maps rarely show the French names outside of Quebec, which itself is spelled *Québec* in French.

The study of placenames is called toponymy, while that of the origin and historical usage of placenames as words is etymology.

In order to improve legibility or to aid the illiterate, some maps have been produced using pictograms to represent places. The iconic example of this practice is Lance Wyman's early plans for the Mexico City Metro, on which stations were shown simply as stylized logos. Wyman also prototyped such a map for the Washington Metro, though ultimately the idea was rejected. Other cities experimenting with such maps are Fukuoka, Guadalajara and Monterrey.

Map Symbology

Cartographic symbology encodes information on the map in ways intended to convey information to the map reader efficiently, taking into consideration the limited space on the map, models of human understanding through visual means, and the likely cultural background and education of the map reader. Symbology may be implicit, using universal elements of design, or may be more specific to cartography or even to the map.

A map may have any of many kinds of symbolization. Some examples are:

- A legend, or key, explains the map's pictorial language.
- A title indicates the region and perhaps the theme that the map portrays.
- A neatline frames the entire map image.
- A compass rose or north arrow provides orientation.
- An overview map gives global context for the primary map.
- A bar scale translates between map measurements and real distances.
- A map projection provides a way to represent the curved surface on the plane of the map.

The map may declare its sources, accuracy, publication date and authorship, and so forth. The map image itself portrays the region.

Map coloring is another form of symbology, one whose importance can reach beyond aesthetic. In complex thematic maps, for example, the color scheme's structure can critically affect the reader's ability to understand the map's information. Modern computer displays and print technologies can reproduce much of the gamut that humans can perceive, allowing for intricate exploitation of human visual discrimination in order to convey detailed information.

Quantitative symbols give a visual indication of the magnitude of the phenomenon that the symbol represents. Two major classes of symbols are used to portray quantity. Proportional symbols change size according to phenomenon's magnitude, making them appropriate for representing statistics. Choropleth maps portray data collection areas, such as counties or census tracts, with color. Using color this way, the darkness and intensity (or value) of the color is evaluated by the eye as a measure of intensity or concentration.

Map Key or Legend

Legend or key of a Belgian road map

The map key, or legend, describes how to interpret the map's symbols and may give details of publication and authorship.

Examples of Point Symbols

Symbol	Explanation
	mine (Hammer and pick symbol), former mine
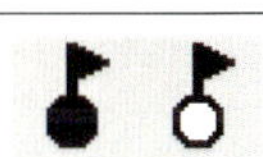	castle, Burg
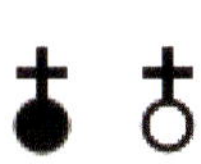	church, chapel, monastery (♁)

	monument
	Hotel
	airport
	railway station
	Tourist information

Map Generalization

A good map has to compromise between portraying the items of interest (or themes) in the right place on the map, and the need to show that item using text or a symbol, which take up space on the map and might displace some other item of information. The cartographer is thus constantly making judgements about what to include, what to leave out and what to show in a *slightly* incorrect place. This issue assumes more importance as the scale of the map gets smaller (i.e. the map shows a larger area) because the information shown on the map takes up more space *on the ground*. A good example from the late 1980s was the Ordnance Survey's first digital maps, where the *absolute* positions of major roads were sometimes a scale distance of hundreds of metres away from ground truth, when shown on digital maps at scales of 1:250,000 and 1:625,000, because of the overriding need to annotate the features.

Map Projections

The Earth being spherical, any flat representation generates distortions such that shapes and areas cannot both be conserved simultaneously, and distances can never all be preserved. The mapmaker must choose a suitable *map projection* according to the space to be mapped and the purpose of the map.

Cartographic Errors

Some maps contain deliberate errors or distortions, either as propaganda or as a "watermark" to help the copyright owner identify infringement if the error appears in competitors' maps. The latter often come in the form of nonexistent, misnamed, or misspelled "trap streets". Other names and forms for this are paper townsites, fictitious entries, and copyright easter eggs.

Another motive for deliberate errors is cartographic "vandalism": a mapmaker wishing to leave his or her mark on the work. Mount Richard, for example, was a fictitious peak on the Rocky Mountains' continental divide that appeared on a Boulder County, Colorado map in the early 1970s. It is believed to be the work of draftsman Richard Ciacci. The fiction was not discovered until two years later.

Sandy Island (New Caledonia) is an example of a fictitious location that stubbornly survives, reappearing on new maps copied from older maps while being deleted from other new editions.

Terrain Cartography

The depiction of the land surface features have been a challenge for map makers ever since the earliest maps.Some of the oldest maps show mountains as rounded `mole hill' or triangular shapes as they might appear if one were looking at them from the level of a valley. The plan view that we have been trained to interpret was unknown to the early cartographer and the sophisticated survey methods that would provide elevation values for the construction of a measured plan view were far away into the future.

The simple rounded dome shape which resembles a mole hill was the most common early cartographic form of a mountain.These simple shapes were often arranged, even in the earliest maps in row or chains, and often aligned with the axis of the row. In poor copies of these maps rows of mountain symbols became bands and ridges. The row arrangement in Figure 3b went out of use during the 15th century and the instead of the symbols having varied orientations they all began to line up with the line of sight of the viewer. Groupings of the symbols were created following the upright rows of symbols and are described as the fish scale representation. Further modifications included outlines of cone-shaped, and craggy pointed mountains which became differentiated by size. Shadowing was suggested with lines and the imaginary light source was usually from the left.

In the 16th century demands on the cartographer increased with the spiritual and intellectual awakening of the Renaissance. Topographic survey methods were evolving to include the compass, measuring chain and measuring cart which provided more accurate metrics for the production of maps. Artists and scientists also became more interested in the natural world and produced maps such as Figure by Leonardo da Vinci. About 50 years later cartographers developed more natural symbology for mountains following da Vinci's example. The previously flattened landscape with isolated mountain symbols gave way to a continuous terrain of differentiated graphical symbology. Forests and fields supported the oblique view of continuous terrain and three dimensionality was achieved through slope lines and the use of shadow hachures.

Another technological development of the 16th century provided finer, smoother and denser lines. The earliest printed maps were produced using the woodcut, but the technique of copper engraving allowed for an increase of topographic content and the character of maps were changed due to the technique. Unfortunately the increases in content were often accompanied by a reduction in drawing skill.

Continued developments in survey techniques during the 17th and 18th century prompted further advancements in cartographic presentation. It was an uneven development as both the technological and artist aspects moved along in separate irregular paths. A noteworthy example of the heights to which the technological and artistic aspects of cartography could attain are the maps by Hans Conrad Gyger during the mid 17th century. Gyger combined detailed, accurate surveying with a presentation in plan view. The entire map, of which Figure is a small portion, was nearly 5 meters square, naturally colored, and shows excellent 3 dimensional realism. This is the earliest relief map in existence but it did not have an effect on the cartography of the time because it was hidden away as a military secret.

Only in the 18th century did real advances such as those seen in Gyger's map begin to appear. France was the leader in cartography with the first modern national triangulation completed in 1774 by Cassini and the advances in geodetic survey and calculation an copper engraving were evident in the maps produced by Cassini and his son. Although the precise planimetric nature of the details could have contributed to advancements in terrain representation the Cassini maps did not. During the 18th century a wide range of cartographic developments coexisted. The transition to a planimetric view took place near the turn of the century. Although maps were now being created with a planimetric view there was still no means for cartographers of the time to actually see what mountains looked like from above. The restriction imposed by copper engraving still was limiting the drawings essentially to line work in one color. Areas of slope were depicted hachures with the level areas left blank. Hachures were drawn with the rule that the greater the slope and ruggedness, the denser and darker were the hachures. Frederick the Great was quoted as instructing his Prussian military topographers, "Wherever I can't go, let there be a blot". This approach was carried to such an extreme that the convention "the steeper, the thicker and darker the pattern of hachuring" replaced the realistic three dimensional typical style of Gyger.

In 1799 the Saxon military topographer, Johann Georg Lehmann, developed a system called slope hachuring. Achure lines are drawn in the direction of slope with the hachures broken into nearly even sections. The hachure were arranged so that their end points began and ended on what were assumed to be contour lines with regular vertical intervals. Flatter terrain were drawn with longer hachures and steeper slopes with shorter. The angle of the slope determined the thickness of the hachures. Steep slopes had heavier hachures and gentle slopes had finer. This system allowed for easy identification of the direction of slope and the change from ground level to steep slope and cartographers could produce the hachures in an objective and consistent manner. Slope hachuring became the most means of showing terrain through the 19th century up until the middle of the 20th century.

Meanwhile, in France, Switzerland and a bit in Italy hachuring was also developed but with a tendency towards the use of "left illumination". Heavier hachures were drawn on the shaded slopes and finer hachures were drawn on the illuminated sides. Building upon this was the combination of using Lehmann's arrangements of hachure lines with the three dimensional effect provided by the "left illumination" method. The result is called shadow hachure. The Dufour map seen in Figure is the "finest and clearest map of any high mountain region to have appeared in the last century." This method was in use until the middle of the 20th century for many national and private maps.

While the debate over the relative merits of the two methods of hachuring were being debated a new method emerged which would come to dominate the depiction of terrain. The new method was the use of contours and depth contours or isobaths. Contours have the immense advantage over other methods of terrain depiction because they record and convey the geometric form of the terrain including elevations and the angles and directions of slopes. The earliest example of isobaths, or depth contours, was by Pieter Buuinss in 1584. Contours were known and used, mainly in the form of isobaths, while hachure techniques were being developed and used. Contour maps of the land surface had to wait until advances in survey techniques provided the detail required for the production of these maps.

During the past 100 years the demands for content and accuracy have increased until the classical survey methods became inadequate. Photogrammetric methods were developed and these provided the increase in speed and accuracy required for contour maps. These methods have also evolved to make use of satellite imagery as the source of information about the land surface.

Even before these advances, the development of lithography in 1796 and its application to printing maps in 1826 changed the appearance of maps. Multicolored maps were now possible. First hachures were changed to brown while other elements such as contours or different land surfaces could be printed in various colors. Hachures evolved into shading tones, or shadows and areas could now be printed in continuous color. During the latter part of the 19th century maps appeared using regional area coloring as hypsometric tints and naturalistic and symbolic landscape tints. Advances of the recent past 100 years are familiar to us in the form of multicolored maps, standardized use of contours for topographic map series, and hill shaded effects. The older method of hachures has essentially disappeared. Those methods currently in use for the depiction of terrain will be explored in more detail in the following sections.

Hill Profiles

From a 1639 map of Hispaniola by Joan Vinckenboons, showing use of hill profiles

The most ancient form of relief depiction in cartography, hill profiles are simply drawings of mountains and hills in profile, placed as appropriate on generally small-scale (broad area of coverage) maps. They are seldom used today except as part of an "antique" styling.

Hachures

Dufour map of Bern (1907); this is a shaded hachure map.

Hachures are also an older mode of representing relief. They are a form of shading, although different from the one used in shaded maps. They show the orientation of slope, and by their thickness and overall density they provide a general sense of steepness. Being non-numeric, they are less useful to a scientific survey than contours, but can successfully communicate quite specific shapes of terrain.

Hachure representation of relief was standardized by the Austrian topographer Johann Georg Lehmann in 1799.

Contour Lines

Contour lines (or isohypses) are isolines showing equal elevation. This is the most common way of numerically showing elevation, and is familiar from topographic maps.

Most 18th and early 19th century national surveys did not record relief across the entire area of coverage, calculating only spot elevations at survey points. The United States Geological Survey (USGS) topographical survey maps included contour representation of relief, and so maps that show relief, especially with exact representation of elevation, came to be called topographic maps (or "topo" maps) in the United States, and the usage has spread internationally.

Siegfried map of Bernina Pass (1877) with black, blue and brown contour lines at 30-meter intervals

On maps produced by Swisstopo, the color of the contour lines is used to indicate the type of ground: black for bare rock and scree, blue for ice and underwater contours, brown for earth-covered ground.

Tanaka (Relief) Contours

The Tanaka (Relief) Contours technique is a method used to illuminate contour lines in order to help visualize terrain. Lines are highlighted or shaded depending on their relationship to a light source in the Northwest. If the object being illustrated would shadow a section of contour line, that contour would be represented with a black band. Otherwise, slopes facing the light source would be represented by white bands.

This method was developed by Professor Tanaka Kitiro in 1950, but had been experimented with as early as 1870, with little success due to technological limitations in printing. The resulting terrain at this point was a grayscale image. Cartographer Berthold Horn later created software to digitally produce Tanaka Contours, and Patrick Kennelly, another cartographer, later found a way to add color to these maps, making them more realistic.

There are a number of issues with this method. Historically, maps using Tanaka Contours could only be produced on gray backgrounds, which technology was unprepared for. This method is also very time consuming. In addition, the method takes into account mainly aspect and elevation, whereas slope is un accounted for resulting in a very terraced appearance.

Hypsometric Tints

Hypsometric tints (also called layer tinting, elevation tinting, elevation coloring, or hysometric coloring) are colors placed between contour lines to indicate elevation. These tints are shown as bands of color in a graduated scheme or as a color ramp applied to contour lines themselves. A typical scheme progresses from dark greens for lower elevations, up through yellows/browns and on to grays and white at the highest elevations. Hypsometric tinting of maps and globes is often accompanied by a similar method of bathymetric tinting to convey depth of oceans; lighter shades of blue represent shallower water such as the continental shelf and darker shades deeper regions.

Shaded Relief

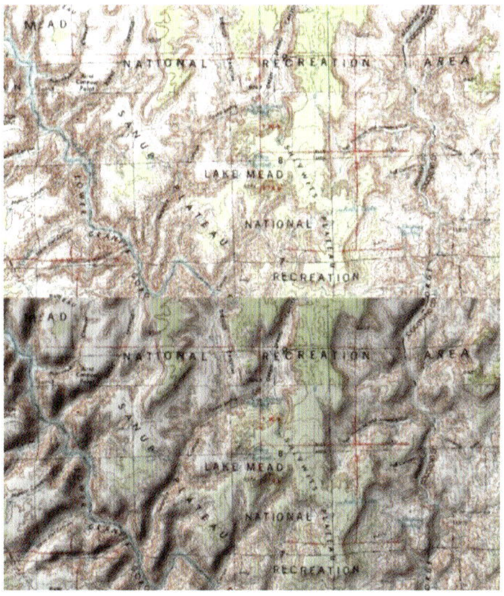

Top: map of Lake Mead area Bottom: the same map with sun shading

Shaded relief, or hill-shading, shows the shape of the terrain in a realistic fashion by showing how the three-dimensional surface would be illuminated from a point light source. The shadows normally follow the convention of top-left lighting in which the light source is placed near the upper-left corner of the map. If the map is oriented with north at the top, the result is that the light appears to come from the north-west. Although this is unrealistic lighting in the northern hemisphere, using a southern light source can cause multistable perception illusions, in which the topography appears inverted.

Shaded relief was traditionally drawn with charcoal, airbrush and other artist's media. The Swiss master Eduard Imhof is widely regarded as the master of manual hill-shading technique and theory.

Imhof's contributions included a multi-color approach to shading, with purples in valleys and yellows on peaks, which is known as "illuminated shading." Illuminating the sides of the terrain facing the light source with yellow colors provides greater realism, enhances the sense of the three-dimensional nature of the terrain, and make the map more aesthetically pleasing and artistic-looking.

Shaded relief is today almost exclusively computer-generated using digital elevation models (DEM), with a resulting different look and feel. Much work has been done in digitally recreating the work of Eduard Imhof, which has been fairly successful in some cases. The DEM may be converted to shaded relief using software such as Photoshop, QGIS, GRASS GIS or ArcMap's Spatial Analyst extension. Computer-generated effects to increase the realism of shaded relief include resolution bumping, the blending of hill-shades at different scales, and sky model shading, which simulates ambient light on terrain from many directions.

Physiographic Illustration

Web version of Patteron's Physical Map of the Coterminous United States. Note the appearance of the Rocky Mountains in the full size version.

Pioneered by Erwin Raisz, this technique uses generalized texture to imitate landform shapes over a large area. A combination of hill profile and shaded relief, this style of terrain representation is simultaneously idiosyncratic to its creator -- often hand-painted -- and found insightful in illustrating geomorphological patterns.

More recently, Tom Patterson created a computer-generated map of the United States using Erwin Raisz's work as a starting point, the *Physical Map of the Coterminous United States*; he called the technique *plan oblique relief.*

Cartographic Principles & Design

Cartographers apply many design principles when compiling their maps and constructing page layouts. Five of the main design principles are legibility, visual contrast, figure-ground, hierarchical organization, and balance. Together these form a system for seeing and understanding the relative importance of the content in the map and on the page. Without these, map-based communication will fail. Together visual contrast and legibility provide the basis for seeing the contents on the map. Figure-ground, hierarchical organization, and balance lead the map reader through the contents to determine the importance of things and ultimately find patterns.

Visual Contrast

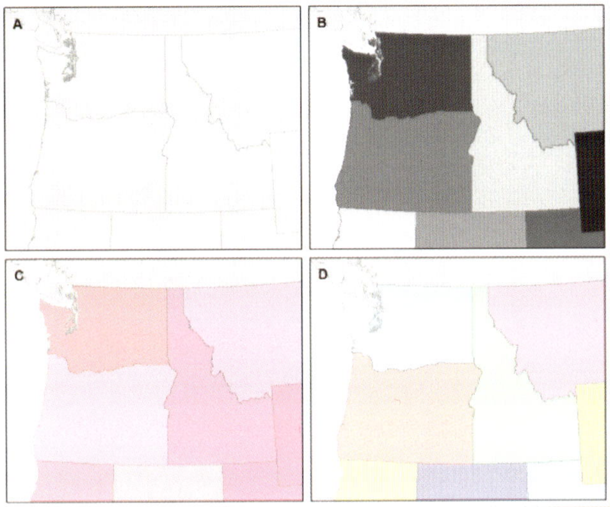

When there is no variation in visual contrast (A), the map reader has a hard time distinguishing features from the background. For quantitative distributions (B), there must be enough contrast between tones for the reader to distinguish unique classes. For qualitative distributions (C),using variations of a single color hue (e.g., red) does not provide as much contrast as using a variety of hues (e.g., red, green, blue, etc.)

Visual contrast which relates to how map features and page elements contrast with each other and their background. To understand this principle at work, consider your inability to see well in a dark environment. Your eyes are not receiving much reflected light so there is little visual contrast between features and you cannot easily distinguish objects from one another or from their surroundings. Add more light and you are now able to contrast features from the background. This concept of visual contrast also applies in cartography. A well-designed map with a high degree of visual contrast can result in a crisp, clean, sharp-looking map. The higher the contrast between features, the more something will stand out, usually the feature that is darker or brighter. Conversely, a map that has low visual contrast can be used to promote a more subtle impression. Features that have less contrast will appear to belong together.

Legibility

Legibility is "the ability to be seen and understood". Many people work to make their map contents and page elements easily seen, but it is also important that they can be understood. Legibility depends on good decision-making for selecting symbols that are familiar and choosing appropriate sizes so that the results are effortlessly seen and easily understood. Geometric symbols are easier to read at smaller sizes; more complex symbols require larger amounts of space to be legible.

Visual contrast and legibility are the basis for seeing. In addition to being able to distinguish features from one another and the background, the features need to be large enough to be seen and to be understood in order for your mind to decipher what you eyes are detecting. Visual contrast and legibility can also be used to promote the other design principles: figure-ground, hierarchical organization, and balance.

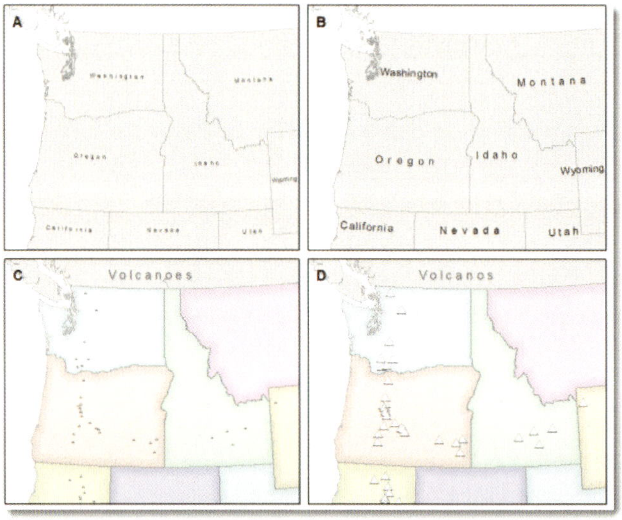

Text and symbols (A and C) that are too small cannot be seen. Once able to be seen (B and D), they must also be understood.

Figure-ground

Figure-ground organization is the spontaneous separation of the figure in the foreground from an "amorphous" background. Cartographers use this design principle to help their map readers find the area of the map or page to focus on. There are many to promote figure-ground organization, such as adding detail to the map or using a white wash, a drop shadow, or feathering.

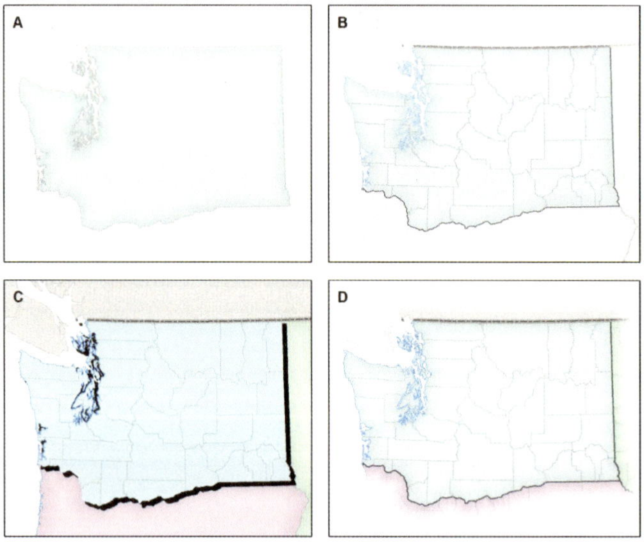

Using closed forms (A), a white wash (B) , a drop shadow (C), or feathering (D) will promote figure-ground organization on your map.

Hierarchical Organization

One of the major objectives in map making is to "separate meaningful characteristics and to portray likenesses, differences, and interrelationships" (Robinson, et al,. 1995, p. 327). The internal graphic structuring of the map (and the page layout more generally) is fundamental to helping people read your map. You can think of a hierarchy as the visual separation of your map into layers of information. Some types of features will be seen as more important than other kinds of features, and some features will seem more important than other features of the same type. Some page elements (e.g., the map) will seem more important than others (e.g., the title or legend). This visual layering of information within the map and on the page helps readers focus on what is important and enables them to identify patterns. Hierarchical organization on reference maps (those that show the location of a variety of physical and cultural features, such as terrain, roads, boundaries, and settlements) works differently than on thematic maps (maps that concentrate on the distribution of a single attribute or the relationship among several attributes). For reference maps, many of the features should be no more important than one another and so, visually, they should lie on essentially the same visual plane. In reference maps, hierarchy is usually more subtle and the map reader brings elements to the forefront by focusing attention on them. For thematic maps, the theme is more important than the base that provides geographic context.

Balance

Balance involves the organization of the map and other elements on the page. A well-balanced map page results in an impression of equilibrium and harmony. We can also use balance in different ways to promote "edginess" or "tension" or create an impression that is more "organic". Balance results from two primary factors, visual weight and visual direction. If you imagine that the center of your map page is balancing on a fulcrum, the factors that will "tip" the map in a particular direction include the relative location, shape, size, and subject matter of elements on the page.

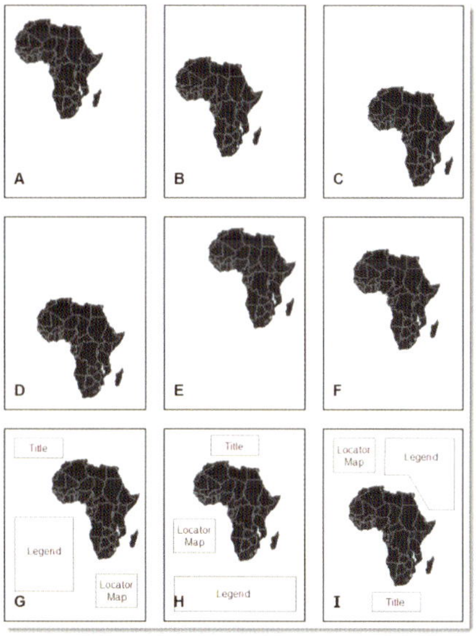

Which of the top six maps seems most balanced? It should appear that (F) has visual equilibrium, usually

achieved by placing the central figure slightly above center on the page. However, the addition of page elements, such as the title and legend, will modify the visual impression, so all content on the page should be evaluated together to judge balance.

Together these five design principles have a significant impact on your map. How they are used will either draw the attention of your map readers or potentially repel them. Giving careful thought to the design of your maps using these principles will help you to assure that your maps are ones people will want to look at!

Map Layout

There are certain points which are to be kept in mind while designing a map:

a. Objective: It is necessary to pay attention to the question why and where will the map going to be used. One should be clear about whether the map is just sharing information, depicting result of an analysis or highlighting the key issues and relationships. One should also know whether the map is going to be displayed on a wall, in a book or in some other information resource.

b. Audience: A designer/cartographer must know who will be addressed through the maps. Maps must match the level/expectations of the audience. The level of design for a technical group and for general people ought to be different.

c. Balanced design: A map must be prepared using appropriate page size, color, patterns of shading, text, and scale. Everything on the map should be legible.

Designing the map layout

The process of map composition starts with preparing a layout for the map. Apart from the data, a map has certain other things that make map a package of effective and clear communication. These provide critical information to users and are known as map elements. A layout specifies the space and positions for different map elements such as neat lines, title, North arrow, scale bar, legend etc. Preparing an effective layout often requires experimentation with the available space.

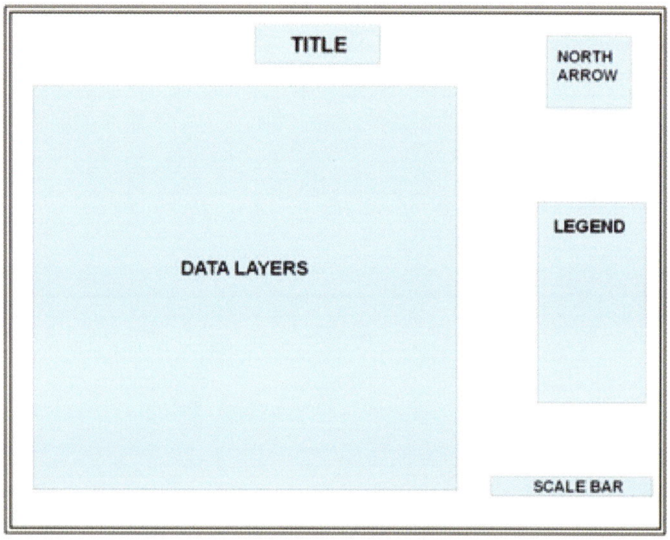

A map Layout

Every element in the layout has to be given a thought—whether it is important to include the respective element, does it require elaboration etc. The map elements that are generally found on a map are:

- Scale: Scale must be given in order to derive the actual size of an entity on the map or distance between two geographical entities on a map.

- Direction: True north is the direction of North Pole and it differs from the magnetic north. The magnetic north pole changes due to the changes in the geo-physical condition of the earth. Many maps indicate both the true and the magnetic north but the direction that is indicated on most of the maps is the true north.
- Legend: Legend lists all the symbols used in a map and describes what they depict.
- Title: A short suitable text that clearly defines the theme of the map.

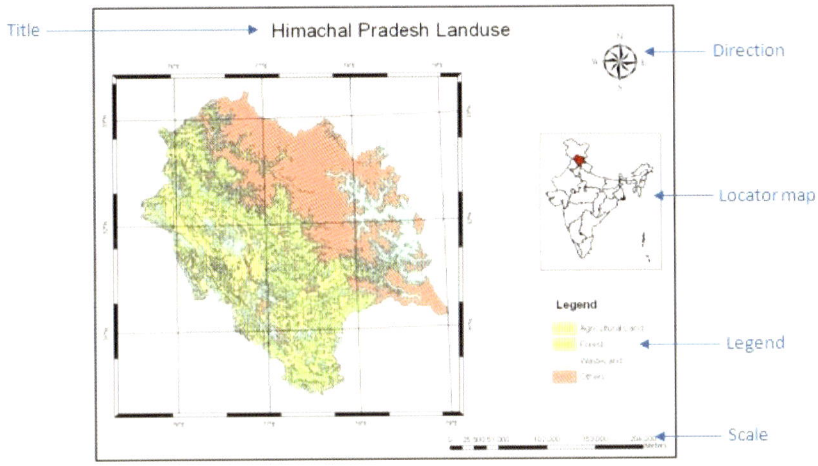

A map with its elements

Given below are a few other elements that are selectively used:

- Neatlines: These are lines used to frame a map to indicate where the map begins and where it ends.
- Reference grid/Graticule: A reference grid is a network of evenly spaced horizontal and vertical lines used to create context on a map. The grid can be used to show unique locations (control points) either in a geographic coordinate system or in a projected coordinate system.
- Locator map: The maps that display locations unfamiliar to users, are generally supported with locator maps which represent the locations in a larger geographic context and with which the user is expected to be familiar.
- Inset map: Sometimes the details on a part of a map are so clustered that they become difficult to read and a magnified view of that part of the map is required. These magnified view maps or close up maps are called insets.
- Source of information: The age, accuracy and reliability of the data sources are critical in carrying out any study. One can show the sources of data one has used in maps so that a user can track them and check his analysis and interpretation.
- Date of production: The representation of time on the maps is important in some cases. For example the weather map prepared on daily, weekly or monthly basis must indicate time on them. A road map to be used for a developing city must be a recent one so as to be relevant for the city etc.

Map Designing

The aim of filling colors in a map is to make visual distinction among various features thus making map more decipherable.

Hue is the dominant wavelength we usually call as color such as green, red or blue. Saturation is the purity of hue or the dominance of hue. Value measures how dark or light the color is when hue is maintained constant. Changes in hue usually indicate qualitative differences such as different administrative units whereas changes in value and saturation represent quantitative differences such as population density in a country.

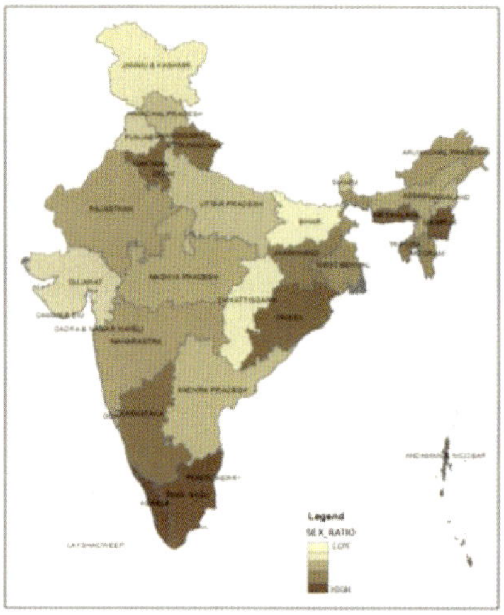

Map showing sex ratio in different states of India

The lightness or darkness of the color represents quantitative differences. Generally, dark colors mean more and represent high values of the attribute under study. The above given map depicts the sex ratio in various states of India. The areas in dark color represent states with high sex ratio and the ones in light color represent states with low sex ratio.

There is a slight difference in color that appears when a map is displayed on a computer screen and when it is printed on a paper. The difference is attributed to the two different ways of creating colors known as additive, and subtractive models of color which are explained below.

Additive Color Model

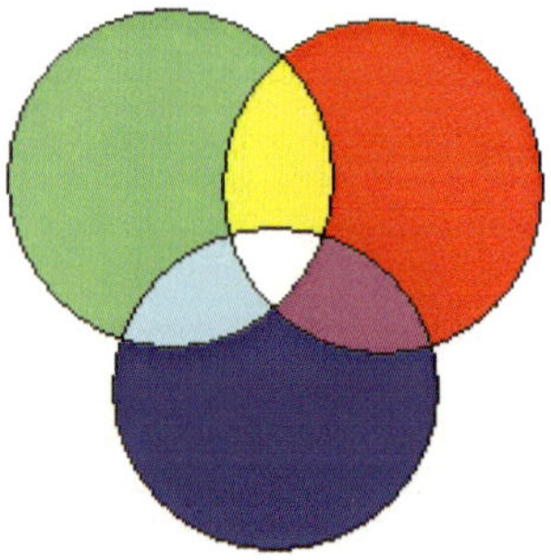

Primary colors
Red, Blue, Green

Secondary colors
Red + Green = Yellow
Red+ Blue = Magenta
Blue+ Green = Cyan

Red+ Green+ Blue = White

Additive color mixing

It involves light emitted from a source and is employed on devices that use light such as camera, monitors etc. The three primary colors red, green, and blue are mixed at different levels to generate other colors. When one of these colors is combined with the other in equal amounts the secondary colors cyan, magenta, and yellow are produced. The combination of all the three colors in equal intensities produces white

Subtractive Color Model

A subtractive color model explains the mixing of dyes, inks etc. to create a range of colors, each formed by subtracting i.e. absorbing some wavelengths of light and reflecting the others. A printer works on this color model. It uses three colors—cyan, magenta and yellow. When these are mixed together black is seem to be formed but the color is actually dark brown. This is why most printers add a black ink to print black text.

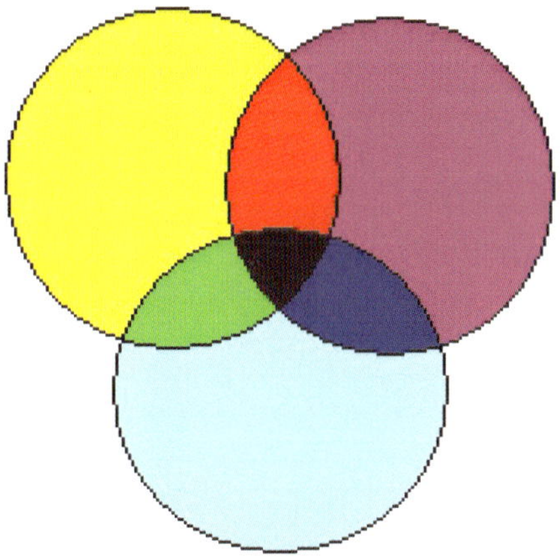

Subtractive color mixing

Text

The use of text on maps enhances the information one wishes to convey. Displaying geographic features and symbols alone on the map doesn't convey the full meaning until and unless it is supported with the relevant text. Inserting text on maps can seve various purposes such as:

- To label the features on map layers, e.g. labeling the name of the states in the political map of India
- To use text as graphic so as to highlight a particular area, e.g. labeling the location of Taj Mahal
- To add information such as title, author, data source references to the map layout.

Placement of Labels

Labeling refers to placing of a descriptive text onto a feature on the map. The placement of labels is an important part of cartographic design because labeling affects the readability of a map. Placing labels at appropriate positions facilitates users to associate labels with the features being described. Sometimes, a part of map becomes overcrowded with labels making it difficult for a user to distinguish which label is used for which feature. This situation can be taken care of in a dynamic map, by adjusting labels in such a manner that they only appear on the map after it is viewed on a particular scale. For static maps, call outs are used for labeling features in such a situation.

Proper placement of labels that mark the dams on Himachal Pradesh map

Placement of labels using call outs that mark lakes in Himachal Pradesh map

Symbols

A symbol is a graphic or a pictorial element used to represent a feature on map. Various types of symbols are used for representing objects or features belonging to any of the three themes viz. point, line and polygon. Every symbol has a set of properties associated with it. These include its shape, size, color, angle, pattern etc

Shape is the geometric form of the symbol. It is used to differentiate between the object classes. The closer the shapes of the symbols resemble the features they represent, the better is the map perceived by the users.

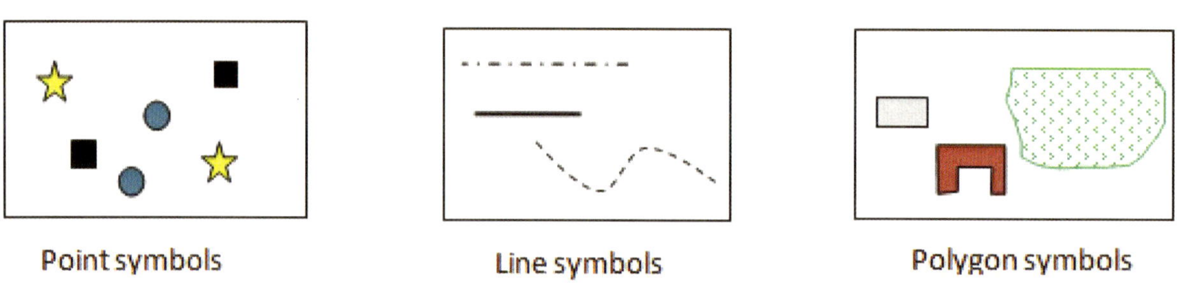

Symbols for different features

The size of symbols depicts a quantitative difference in the distribution of an attribute. Given below is a map showing petrol pump sites in an area. The difference in size of symbols that represent petrol pump stations corresponds to the number of people working in the particular petrol pump station. The biggest sized symbol represents that the station has the highest manpower as compared to the other stations.

Attribute (manpower) representation by size of symbol

In the following map bars have been used to show the comparison and distribution of male and female populations in different states of India. The size of a bar corresponds to the number of individuals (male or female, whatever the bar represents) existing in a particular state.

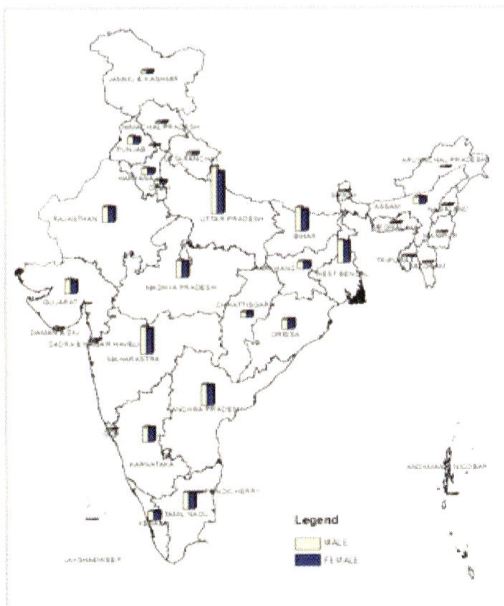

Attribute (male/female population) representation by bar size

Sometimes a combination of symbol and color exhibit the attribute of interest. In the example shown below, the elevation gradation of a landscape is marked by the simultaneous use of contours (lines) and colors.

Toposheet Indexing

For the purpose of an international series (within 4° N to 40° N Latitude and 44° E to 124° E Longitude) at the scale of 1: 1,000,000 is considered as a base map. This map is divided into sections of 4° latitude × 4° longitude and designated from 1 to 136 consisting of the segments that cover only land area.

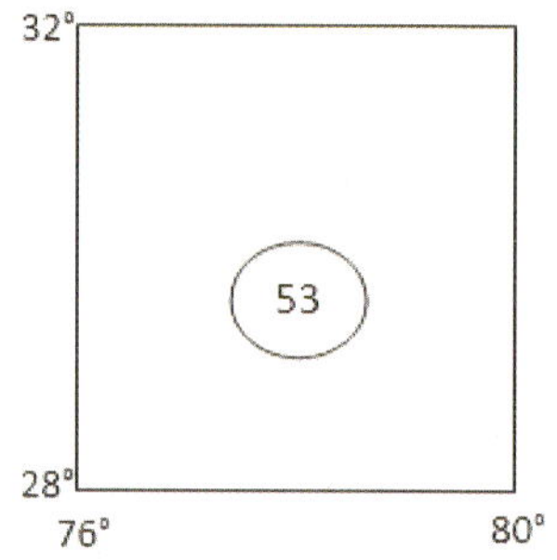

A sheet of 4° × 4° (scale: 1: 1,000,000)

Each section is further divided into 16 sections (4 rows and 4 columns) each of 1° latitude × 1° longitude. The sections start from Northwest direction, run column wise and end in Southeast direction.

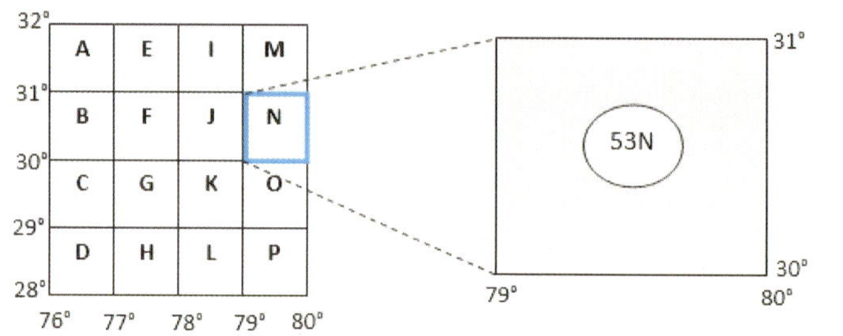

A sheet of 4° × 4° (scale: 1: 1,000,000)

A sheet of 1° × 1° (scale: 1: 250, 000)

The 1°×1° sheets are further subdivided into four parts, each of 30′ latitude × 30′ longitude. These are identified by the cardinal directions NE, NW, SE and SW.

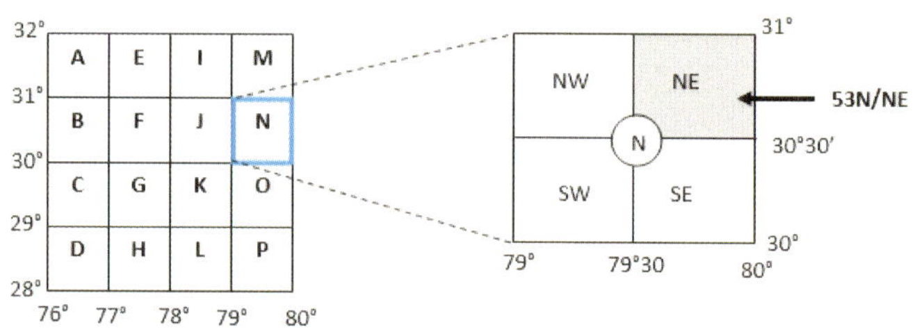

A sheet of 4° × 4° (scale: 1: 1,000,000)

A sheet of 30′ × 30′ (scale: 1:100,000)

The 1°×1° sheets can also be divided into 16 sections each of 15′ latitude × 15′ longitude and are numbered from 1 to 16 in a columned manner.

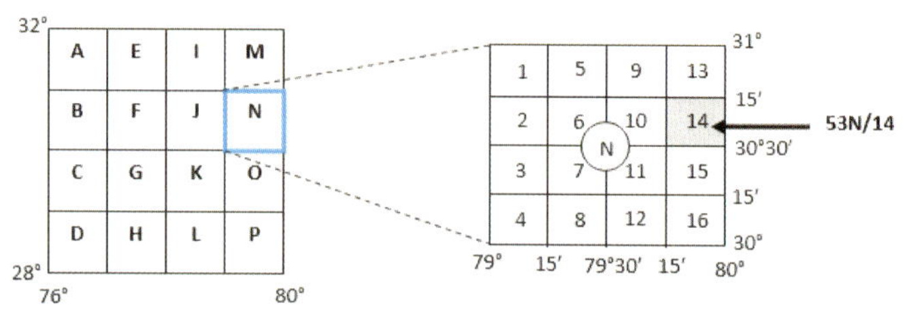

A sheet of 4° × 4° (scale: 1: 1,000,000)

A sheet of 15′ × 15′ (scale: 1: 50,000)

A 15′×15′ sheet can be divided into 4 sheets, each of 7(1/2)′ and are numbered as NW, NE, SW and SE

Map

We see them every day, we use them when we travel, and we refer to them often, but what is a map?

A map is defined as a representation, usually on a flat surface, of a whole or part of an area. The job of a map is to describe spatial relationships of specific features that the map aims to represent. There are many different types of maps that attempt to represent specific things. Maps can display political boundaries, population, physical features, natural resources, roads, climates, elevation (topography), and economic activities.

Maps are produced by cartographers. Cartography refers both the study of maps and the process of map-making. It has evolved from basic drawings of maps to the use of computers and other technologies to assist in making and mass producing maps.

Maps have been around for 8000 years or so. Cartography was significantly advanced in ancient Greece. The notion of a spherical Earth was recognized among Greek philosophers by the time of Aristotle and has been acknowledged by all geographers ever since.

The beginning of map history constitutes from cave painting and rock carvings to ancient maps of Greece, Babylon & Asia, from the age of exploration to the 21st century. People have produced and utilized maps as necessary tools to help them identify, understand and navigate their way around. Mapping has an important step in the human race. They assisted us in our intellectual growth and development and have been passed from generation to generation helping us in the development of our society and culture.

In the Medieval period, European maps were mostly based by religious views. Viking explored the North Atlantic while map-making developed more realistic and practical lines in the Mediterranean region. All maps were drawn by hand, which made the circulation of maps limited.

During the 17th to the 19th century, Maps became more and more accurate and factual with the use of scientific techniques. Various countries adopted national mapping programs. Following World War I, the widespread use of aerial photography helped a lot in map-making process. The combination of ground observations and remote sensing is the base of modern cartography.

Cartography has constantly evolved in order to meet the demands of new generations of map makers' users. The oldest identified maps are from around 2300 B.C. which are preserved on Babylonian clay tablets. The first maps were manually made with brushes on parchments and varied in quality and limited in circulation. The invention of the compass, printing press, telescope, etc allowed the formation of more accurate maps and the ability to make accurate reproductions.

Advances in technologies, has made possible the creation of maps with fine details, no distortion in shape and less resistant to wear and tear. This also removed the need for engraving which reduced the time to create and reproduce maps. Advances in electronic technology in the late 20th century, led to a new revolution in cartography. Computer hardware devices such as computer screens, printers and scanners along with image processing, visualization, spatial analysis and database software, have extended the ability to produce maps that show different features.

Is a Globe a Map?

A globe is a map. Globes are some of the most accurate maps that exist. This is because the earth is a three-dimensional object that is close to spherical. A globe is an accurate representation of the spherical shape of the world. Maps lose their accuracy because they are actually projections of a part of or the entire Earth.

Sometimes a globe shows the topography elevations otherwise they are plain round, some globes include imprinted lines with parallels and meridians for finding the coordinates of a particular place. Globes offer the most precise view of Earth at present.

Advantages of maps over globes:

- Maps are more compact
- Easy to store
- Can hold different range of scales
- Easily viewed on different sized screens
- Can show larger portions of an area at once
- Cheaper to produce and transport.

Uses of Map

A map is useful for both a layman and an intelligent person, as maps contain loads of information. It is up to an individual how he makes use of it. Maps are generally used for:

- Analysis
- Confirmation
- Communication
- Decoration
- Collection
- Investment
- Exploration
- Hypothesis Stimulation
- Navigation
- Control & Planning
- Map Reading
- Storage of Information
- Historical perspective

Geographic Maps

A celestial map from the 17th century, by the cartographer Frederik de Wit

Cartography or *map-making* is the study and practice of crafting representations of the Earth upon a flat surface, and one who makes maps is called a cartographer.

Road maps are perhaps the most widely used maps today, and form a subset of navigational maps, which also include aeronautical and nautical charts, railroad network maps, and hiking and bicycling maps. In terms of quantity, the largest number of drawn map sheets is probably made up by local surveys, carried out by municipalities, utilities, tax assessors, emergency services providers, and other local agencies. Many national surveying projects have been carried out by the military, such as the British Ordnance Survey: a civilian government agency, internationally renowned for its comprehensively detailed work.

In addition to location information maps may also be used to portray contour lines indicating constant values of elevation, temperature, rainfall, etc.

Orientation of Maps

Hereford Cathedral, England, is a classic "T-O" map with
Jerusalem at centre, east toward the top, Europe the bottom left and Africa on the right.

The orientation of a map is the relationship between the directions on the map and the corresponding compass directions in reality. The word "orient" is derived from Latin *oriens*, meaning east. In the Middle Ages many maps, including the T and O maps, were drawn with east at the top (meaning that the direction "up" on the map corresponds to East on the compass). The most common cartographic convention is that north is at the top of a map.

Maps not oriented with north at the top:

- Maps from non-Western traditions are oriented a variety of ways. Old maps of Edo show the Japanese imperial palace as the "top", but also at the centre, of the map. Labels on the map are oriented in such a way that you cannot read them properly unless you put the imperial palace above your head.
- Medieval European T and O maps such as the Hereford Mappa Mundi were centred on Jerusalem with East at the top. Indeed, prior to the reintroduction of Ptolemy's Geography to Europe around 1400, there was no single convention in the West. Portolan charts, for example, are oriented to the shores they describe.
- Maps of cities bordering a sea are often conventionally oriented with the sea at the top.
- Route and channel maps have traditionally been oriented to the road or waterway they describe.
- Polar maps of the Arctic or Antarctic regions are conventionally centred on the pole; the direction North would be towards or away from the centre of the map, respectively. Typical maps of the Arctic have 0° meridian towards the bottom of the page; maps of the Antarctic have the 0° meridian towards the top of the page.
- Reversed maps, also known as Upside-Down maps or South-Up maps, reverse the North is up convention and have south at the top.
- Buckminster Fuller's Dymaxion maps are based on a projection of the Earth's sphere onto an icosahedron. The resulting triangular pieces may be arranged in any order or orientation.
- Modern digital GIS maps such as ArcMap typically project north at the top of the map, but use math degrees (0 is east, degrees increase counter-clockwise), rather than compass degrees (0 is north, degrees increase clockwise) for orientation of transects. Compass decimal degrees can be converted to math degrees by subtracting them from 450; if the answer is greater than 360, subtract 360.

Scale and Accuracy

A 'global view map' of Europe, Western Asia and Africa.

Many maps are drawn to a scale expressed as a ratio, such as 1:10,000, which means that 1 unit of measurement on the map corresponds to 10,000 of that same unit on the ground. The scale statement can be accurate when the region mapped is small enough for the curvature of the Earth to be neglected, such as a city map. Mapping larger regions, where curvature cannot be ignored, requires projections to map from the curved surface of the Earth to the plane. The impossibility of flattening the sphere to the plane without distortion means that the map cannot have constant scale. Rather, on most projections the best that can be attained is accurate scale along one or two paths on the projection. Because scale differs everywhere, it can only be measured meaningfully as point scale per location. Most maps strive to keep point scale variation within narrow bounds. Although the scale statement is nominal it is usually accurate enough for most purposes unless the map covers a large fraction of the earth. At the scope of a world map, scale as a single number is practically meaningless throughout most of the map. Instead, it usually refers to the scale along the equator.

Large scale maps, (e.g. 1:10,000), cover relatively small regions in great detail and small scale maps, (e.g. 1:10,000,000), cover large regions such as nations, continents and the whole globe. The large/small terminology arose from the practice of writing scales as numerical fractions: 1/10,000 is larger than 1/10,000,000. There is no exact dividing line between large and small but 1/100,000 might well be considered as a medium scale. Examples of large scale maps are the 1:25,000 maps produced for hikers; on the other hand maps intended for motorists at 1:250,000 or 1:1,000,000 are small scale.

It is important to recognize that even the most accurate maps sacrifice a certain amount of accuracy in scale to deliver a greater visual usefulness to its user. For example, the width of roads and small streams are exaggerated when they are too narrow to be shown on the map at true scale; that is, on a printed map they would be narrower than could be perceived by the naked eye. The same applies to computer maps where the smallest unit is the pixel. A narrow stream say must be shown to have the width of a pixel even if at the map scale it would be a small fraction of the pixel width.

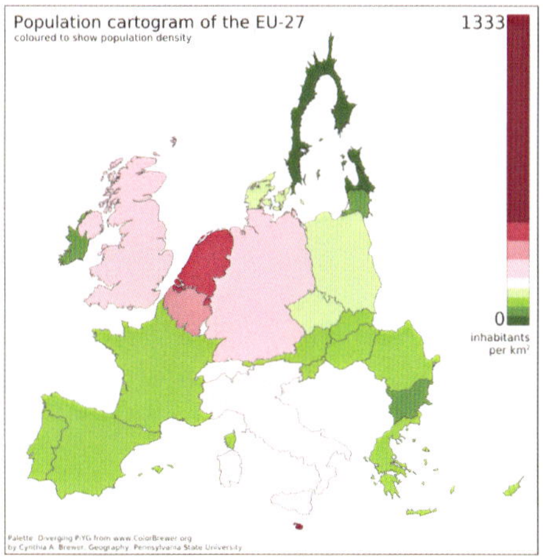

Cartogram: The EU distorted to show population distributions.

Some maps, called cartograms, have the scale deliberately distorted to reflect information other than land area or distance. For example, this map (at the right) of Europe has been distorted to show population distribution, while the rough shape of the continent is still discernible.

Another example of distorted scale is the famous London Underground map. The basic geographical structure is respected but the tube lines (and the River Thames) are smoothed to clarify the relationships between stations. Near the center of the map stations are spaced out more than near the edges of map.

Further inaccuracies may be deliberate. For example, cartographers may simply omit military installations or remove features solely in order to enhance the clarity of the map. For example, a road map may not show railroads, smaller waterways or other prominent non-road objects, and even if it does, it may show them less clearly (e.g. dashed or dotted lines/outlines) than the main roads. Known as decluttering, the practice makes the subject matter that the user is interested in easier to read, usually without sacrificing overall accuracy. Software-based maps often allow the user to toggle decluttering between ON, OFF and AUTO as needed. In AUTO the degree of decluttering is adjusted as the user changes the scale being displayed.

Map Types and Projections

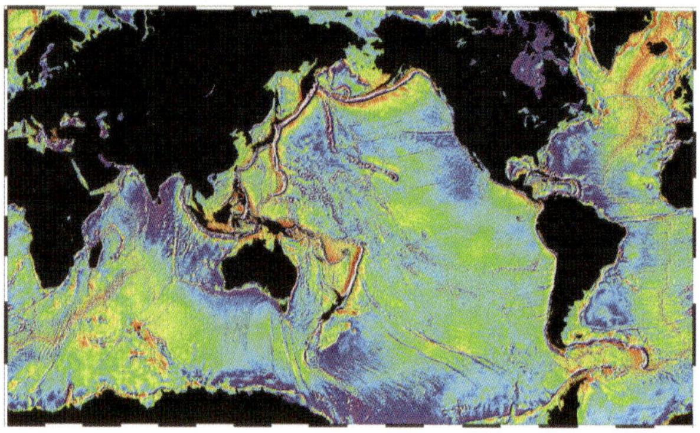

Map of large underwater features.

Maps of the world or large areas are often either 'political' or 'physical'. The most important purpose of the political map is to show territorial borders; the purpose of the physical is to show features of geography such as mountains, soil type or land use including infrastructure such as roads, railroads and buildings. Topographic maps show elevations and relief with contour lines or shading. Geological maps show not only the physical surface, but characteristics of the underlying rock, fault lines, and subsurface structures. Maps that depict the surface of the Earth also use a projection, a way of translating the three-dimensional real surface of the geoid to a two-dimensional picture. Perhaps the best-known world-map projection is the Mercator projection, originally designed as a form of nautical chart. Aeroplane pilots use aeronautical charts based on a Lambert conformal conic projection, in which a cone is laid over the section of the earth to be mapped. The cone intersects the sphere (the earth) at one or two parallels which are chosen as standard lines. This allows the pilots to plot a great-circle route approximation on a flat, two-dimensional chart.

- Azimuthal or Gnomonic map projections are often used in planning air routes due to their ability to represent great circles as straight lines.
- Richard Edes Harrison produced a striking series of maps during and after World War II for Fortune magazine. These used "bird's eye" projections to emphasise globally strategic "fronts" in the air age, pointing out proximities and barriers not apparent on a conventional rectangular projection of the world.

Electronic Maps

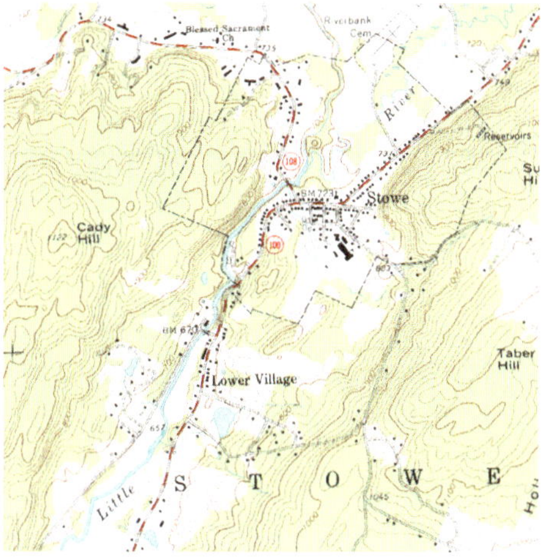

A USGS digital raster graphic.

From the last quarter of the 20th century, the indispensable tool of the cartographer has been the computer. Much of cartography, especially at the data-gathering survey level, has been subsumed by Geographic Information Systems (GIS). The functionality of maps has been greatly advanced by technology simplifying the superimposition of spatially located variables onto existing geographical maps. Having local information such as rainfall level, distribution of wildlife, or demographic data integrated within the map allows more efficient analysis and better decision making. In the pre-electronic age such superimposition of data led Dr. John Snow to identify the location of an outbreak of cholera. Today, it is used by agencies of the human kind, as diverse as wildlife conservationists and militaries around the world.

Relief map Sierra Nevada

Even when GIS is not involved, most cartographers now use a variety of computer graphics programs to generate new maps.

Interactive, computerised maps are commercially available, allowing users to *zoom in* or *zoom out* (respectively meaning to increase or decrease the scale), sometimes by replacing one map with another of different scale, centered where possible on the same point. In-car global navigation satellite systems are computerised maps with route-planning and advice facilities which monitor the user's position with the help of satellites. From the computer scientist's point of view, zooming in entails one or a combination of:

1. replacing the map by a more detailed one
2. enlarging the same map without enlarging the pixels, hence showing more detail by removing less information compared to the less detailed version
3. enlarging the same map with the pixels enlarged (replaced by rectangles of pixels); no additional detail is shown, but, depending on the quality of one's vision, possibly more detail can be seen; if a computer display does not show adjacent pixels really separate, but overlapping instead (this does not apply for an LCD, but may apply for a cathode ray tube), then replacing a pixel by a rectangle of pixels does show more detail. A variation of this method is interpolation.

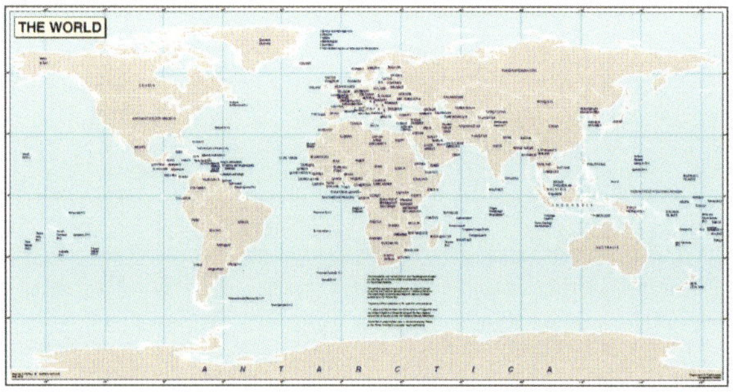

A world map in PDF format.

For example:

- Typically (2) applies to a Portable Document Format (PDF) file or other format based on vector graphics. The increase in detail is limited to the information contained in the file: enlargement of a curve may eventually result in a series of standard geometric figures such as straight lines, arcs of circles or splines.
- (2) may apply to text and (3) to the outline of a map feature such as a forest or building.
- (1) may apply to the text as needed (displaying labels for more features), while (2) applies to the rest of the image. Text is not necessarily enlarged when zooming in. Similarly, a road represented by a double line may or may not become wider when one zooms in.
- The map may also have layers which are partly raster graphics and partly vector graphics. For a single raster graphics image (2) applies until the pixels in the image file correspond to the pixels of the display, thereafter (3) applies.

Climatic Maps

The maps that reflect the territorial distribution of climatic conditions based on the results of long-term observations are climatic maps. Climatic maps can be compiled both for individual climatic features (temperature, precipitation, humidity) and for combinations of them at the earth's surface and in the upper layers of the atmosphere. Climatic maps afford a very convenient overview of the climatic features in a large region and permit values of climatic features to be compared in different parts of the region. Through interpolation the maps can be used to determine the values of climatic features in any particular spot.

Climatic maps generally apply to individual months and to the year as a whole, sometimes to the four seasons, to the growing period, and so forth. On maps compiled from the observations of ground meteorological stations, atmospheric pressure is converted to sea level. Air temperature maps are compiled both from the actual values observed on the surface of the earth and from values converted to sea level. The pressure field in free atmosphere is represented either by maps of the distribution of pressure at different standard altitudes—for example, at every kilometer above sea level—or by maps of baric topography on which altitudes (more precisely geopotentials) of the main isobaric surfaces (for example, 900, 800, and 700 millibars) counted off from sea level are plotted. The temperature, humidity, and wind on aeroclimatic maps may apply either to standard altitudes or to the main isobaric surfaces.

Isolines are drawn on maps of such climatic features as the long-term mean values (of atmospheric pressure, temperature, humidity, total precipitation, and so forth) to connect points with equal values of the feature in question—for example, isobars for pressure, isotherms for temperature, and isohyets for precipitation. Isoamplitudes are drawn on maps of amplitudes (for example, annual amplitudes of air temperature—that is, the differences between the mean temperatures of the warmest and coldest month). Isanomals are drawn on maps of anomalies (for example, deviations of the mean temperature of each place from the mean temperature of the entire latitudinal zone). Isolines of frequency are drawn on maps showing the frequency of a particular phenomenon (for example, annual number of days with a thunderstorm or snow cover). Isochrones are drawn on maps showing the dates of onset of a given phenomenon (for example, the first frost and appearance or disappearance of the snow cover) or the date of a particular value of a meteorological element in the course of a year (for example, passing of the mean daily air temperature through zero). Isolines of the mean numerical value of wind velocity or isotachs are drawn on wind maps (charts); the wind resultants and directions of prevailing winds are indicated by arrows of different length or arrows with different plumes; lines of flow are often drawn. Maps of the zonal and meridional components of wind are frequently compiled for the free atmosphere. Atmospheric pressure and wind are usually combined on climatic maps. Wind roses, curves showing the distribution of other meteorological elements, diagrams of the annual course of elements at individual stations, and the like are also plotted on climatic maps.

Maps of climatic regionalization, that is, division of the earth's surface into climatic zones and regions according to some classification of climates, are a special kind of climatic map.

Climatic maps are often incorporated into climatic atlases of varying geographic range (globe, hemispheres, continents, countries, oceans) or included in comprehensive atlases. Besides general climatic maps, applied climatic maps and atlases have great practical value. Aeroclimatic maps, aeroclimatic atlases, and agroclimatic maps are the most numerous.

Conventional Signs

The various features shown on a map are represented by conventional signs or symbols. For example, colors can be used to indicate a classification of roads. Those signs are usually explained in the margin of the map, or on a separately published characteristic sheet.

Some cartographers prefer to make the map cover practically the entire screen or sheet of paper, leaving no room "outside" the map for information about the map as a whole. These cartographers typically place such information in an otherwise "blank" region "inside" the map—cartouche, map legend, title, compass rose, bar scale, etc. In particular, some maps contain smaller "sub-maps" in otherwise blank regions—often one at a much smaller scale showing the whole globe and where the whole map fits on that globe, and a few showing "regions of interest" at a larger scale in order to show details that wouldn't otherwise fit. Occasionally sub-maps use the same scale as the large map—a few maps of the contiguous United States include a sub-map to the same scale for each of the two non-contiguous states.

Labeling

To communicate spatial information effectively, features such as rivers, lakes, and cities need to be labeled. Over centuries cartographers have developed the art of placing names on even the densest of maps. Text placement or name placement can get mathematically very complex as the number of labels and map density increases. Therefore, text placement is time-consuming and labor-intensive, so cartographers and GIS users have developed automatic label placement to ease this process.

Non-geographical Spatial Maps

Maps exist of the Solar System, and other cosmological features such as star maps. In addition maps of other bodies such as the Moon and other planets are technically not *geo*graphical maps.

Topological Maps

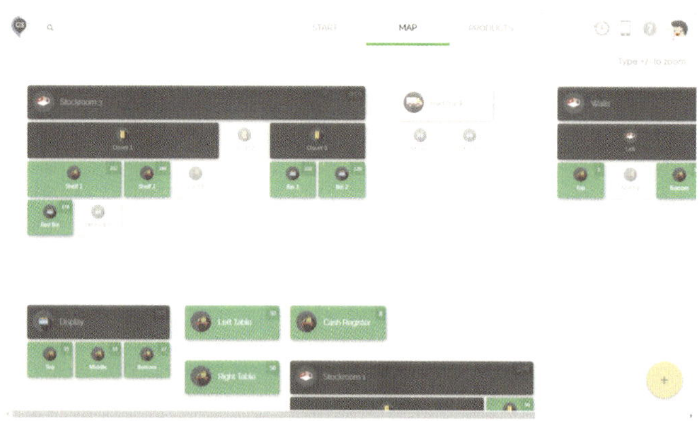

In a topological map, like this one showing inventory locations, the distances between locations is not important. Only the layout and connectivity between them matters.

Diagrams such as schematic diagrams and Gantt charts and treemaps display logical relationships between items, rather than geographical relationships. Topological in nature, only the connectivity is significant. The London Underground map and similar subway maps around the world are a common example of these maps.

General-purpose Maps

General-purpose maps provide many types of information on one map. Most atlas maps, wall maps, and road maps fall into this category. The following are some features that might be shown on general-purpose maps: bodies of water, roads, railway lines, parks, elevations, towns and cities, political boundaries, latitude and longitude, national and provincial parks. These maps give a broad understanding of location and features of an area. The reader may gain an understanding of the type of landscape, the location of urban places, and the location of major transportation routes all at once.

Types of Maps

- Atlas
- Climatic map
- Geologic map
- Physical map
- Political map
- Relief map
- Resource map
- Star map
- Street map
- Thematic map
- Topographical map
- Weather map
- World map
- Cadastral map

Cartogram

"Erwin Raisz called cartograms 'diagrammatic maps.' Today they might be called cartograms, value-by-area maps, anamorphated images or simply spatial transformations. Whatever their name, cartograms are unique representations of geographical space. Examined more closely, the value-by-area mapping technique encodes the mapped data in a simple and efficient manner with no data generalization or loss of detail. Two forms, contiguous and non-contiguous, have become popular. Mapping requirements include the preservation of shape, orientation contiguity, and data that have suitable variation. Successful communication depends on how well the map reader recognizes the shapes of the internal enumeration units, the accuracy of estimating these areas, and effective legend design. Complex forms include the two-variable map. Cartogram construction may be by manual or computer means. In either method, a careful examination of the logic behind the use of the cartogram must first be undertaken."

A cartogram is a type of graphic that depicts attributes of geographic objects as the object's area. Because a cartogram does not depict geographic space, but rather changes the size of objects depending on a certain attribute, a cartogram is not a true map. Cartograms vary on their degree in which geographic space is changed; some appear very similar to a map, however some look nothing like a map at all. There are three main types of cartograms, each have a very different way of showing attributes of geographic objects- Non-contiguous, Contiguous and Dorling cartograms.

Non-contiguous Cartograms

A non-contiguous cartogram is the simplest and easiest type of cartogram to make. In a non-contiguous cartogram, the geographic objects do not have to maintain connectivity with their adjacent objects. This connectivity is called topology. By freeing the objects from their adjacent objects, they can grow or shrink in size and still maintain their shape. Here is an example of two non-contiguous cartograms of population in California's counties.

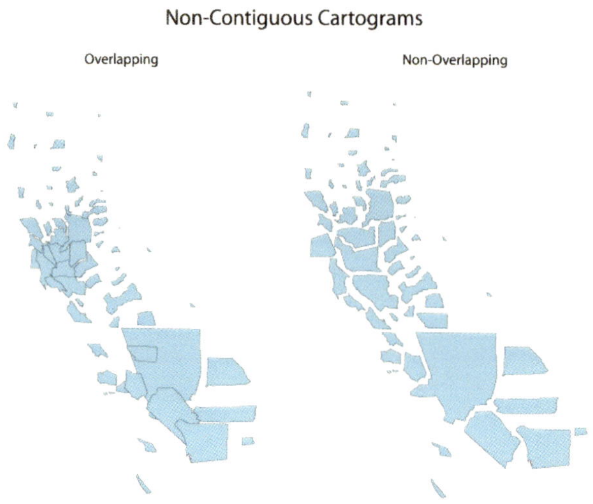

The difference between these two types of non-contiguous cartograms is a significant one. The cartogram on the left has maintained the object's centroid (a centroid is the weighted center point of an area object.) Because the object's center is staying in the same place, some of the objects will begin to overlap when the objects grow or shrink depending on the attribute (in this case population.) In the cartogram on the right, the objects not only shrink or grow, but they also will move one way or another to avoid overlapping with another object. Although this does cause some distortion in distance, most prefer this type of non-contiguous cartogram. By not allowing objects to overlap, the depicted sizes of the objects are better seen, and can more easily be interpreted as some attribute value.

Contiguous Cartograms

In the previous section we referred to the connectivity between objects, or topology. In a non-contiguous cartogram topology was sacrificed in order to preserve shape. In a contiguous cartogram, the reverse is true- topology is maintained (the objects remain connected with each other) but this causes great distortion in shape.

This leads to the single most difficult, but intriguing problem in creating cartograms. The cartographer must make the objects the appropriate size to represent the attribute value, but he or she must also maintain the shape of objects as best as possible, so that the cartogram can be easily interpreted. Here is an example of a contiguous cartogram of population in California's counties. Compare this to the previous non-contiguous cartogram.

Contiguous Cartogram

Dorling Cartogarms

This type of cartogram was named after its inventor, Danny Dorling of the University of Leeds. A Dorling cartogram maintains neither shape, topology nor object centroids, though it has proven to be a very effective cartogram method. To create a Dorling cartogram, instead of enlarging or shrinking the objects themselves, the cartographer will replace the objects with a uniform shape, usually a circle, of the appropriate size. Professor Dorling, for the reason described above in the non-contiguous cartogram section, suggests that the shapes not overlap but rather be moved so that the full area of each shape can be seen. Below is an example of a Dorling cartogram, using the same population of California counties example. Another Dorling-like cartogram is the Demers Cartogram, which is different in two ways. It uses squares rather than circles; this leaves fewer gaps between the shapes. Secondly, the Dorling Cartogram attempts to move the figures the shortest distance away from their true locations; the Demers cartogram often sacrifices distance to maintain contiguity between figures, and it will also sacrifice distance to maintain certain visual cues (The gap between figures used to represent San Francisco Bay in the Demers Cartogram below is a good example of a visual cue.) The 25 Most Populated Counties in California are labeled in each of the two cartograms below for reference.

Dorling and Dorling-like Cartograms

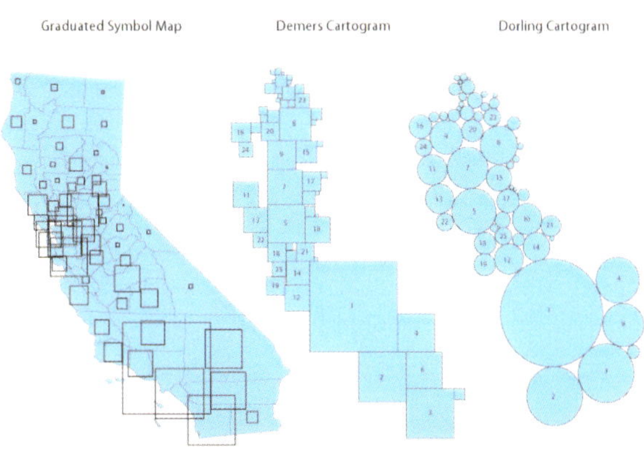

Pseudo-cartograms

Pseudo-cartograms (or false cartograms) are representations that may look like cartograms but do not follow certain cartogram rules. Perhaps the most famous type of pseudo-cartogram was developed by Dr. Waldo Tobler. In this case, instead of enlarging or shrinking the objects themselves, Tobler moves the object's connections to a reference grid such as latitude or longitude in order to give the same effect. This maintains good directional accuracy in the cartogram (if county A is directly north of county B, it will still remain directly north in the cartogram. Note in previous examples, such as the Dorling Cartogram, this is not always true) however, this is a false cartogram because it creates extensive error in the actual size of the objects. Let us consider this error where it becomes very clear, in the case of the California population. Mono County is a very lowly populated county, with only 13,000 people but by pure coincidence it lays at the same latitude as San Francisco County (777,000 people) and at the same longitude as Los Angeles County (9,519,000 people.) It would be impossible to expand the latitude and longitude lines to make Los Angeles and San Francisco the appropriate size without in turn, expanding Mono County as well. In the same way, the lines cannot be contracted to make Mono County the appropriate size without making San Francisco and Los Angeles too small.

Area Cartograms

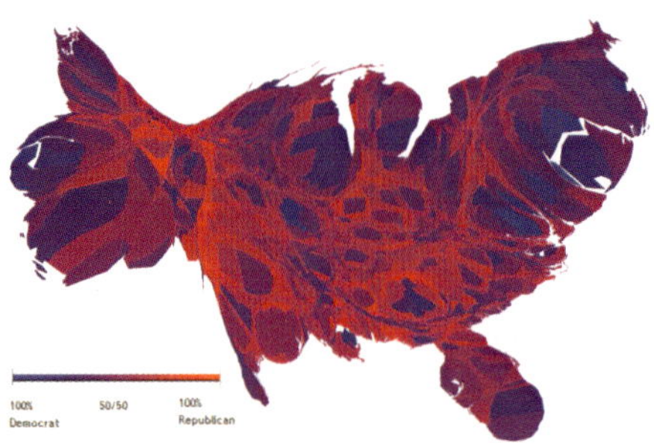

Area cartogram of the United States, with each county rescaled in proportion to its population. Colors refer to the results of the 2004 U.S. presidential election popular vote.

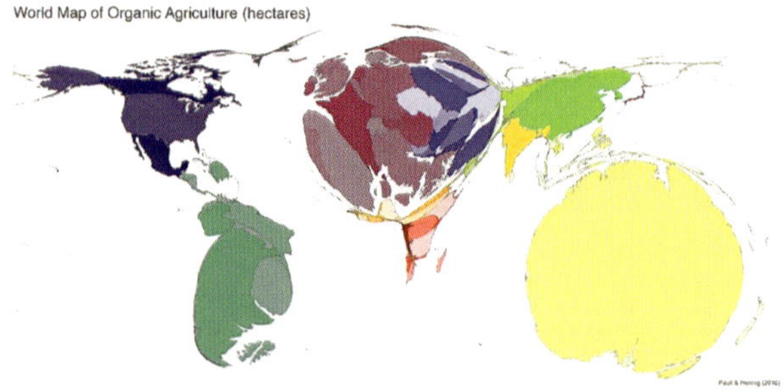

Area cartogram of the world with each country rescaled in proportion to the hectares of certified organic farming

An area cartogram is sometimes referred to as a *value-by-area map* or an *isodemographic map*, the latter particularly for a *population cartogram*, which illustrates the relative sizes of the populations of the countries of the world by scaling the area of each country in proportion to its population; the shape and relative location of each country is retained to as large an extent as possible, but inevitably a large amount of distortion results. Other synonyms in use are *anamorphic map*, *density-equalizing map* and *Gastner map*.

Area cartograms may be contiguous or noncontiguous. The area cartograms shown on this page are all contiguous, while a good example of a noncontiguous cartogram was published in *The New York Times*. This method of cartogram creation is sometimes referred to as *the projector method* or *scaled-down regions*.

Cartograms may be classified also by the properties of shape and topology preservation. Classical area cartograms are typically distorting the shape of spatial units to some degree, but they are strict at preserving correct neighborhood relationships between them. Scaled-down cartograms (from the NY Times example) are strictly shape-preserving. Another branch of cartograms introduced by Dorling, replaces actual shapes with circles scaled according to the mapped feature. Circles are distributed to resemble the original topology. Demers cartogram is a variation of Dorling cartogram, but it uses rectangles instead of circles, and attempts to retain visual cues at the expense of minimum distance. Schematic maps based on quad trees can be seen as non shape-preserving cartograms with some degree of neighborhood preservation.

A collection of contiguous area cartograms is available at Worldmapper, which was started by a collaborative team of researchers at the Universities of Sheffield and Michigan.

Production

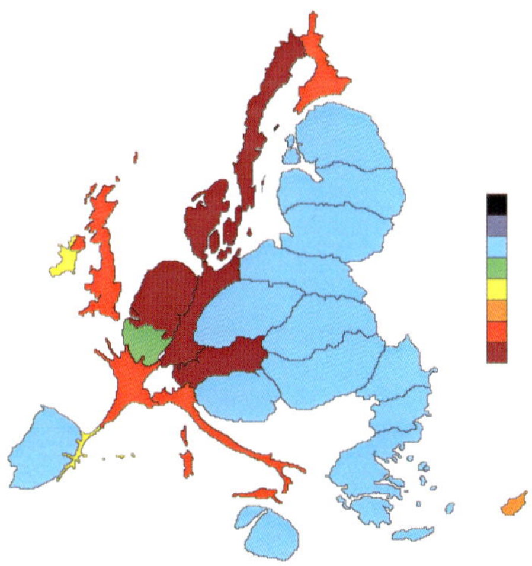

Cartogram showing Open Europe estimate of total European Union net budget expenditure in euros for the whole period 2007-2013, *per capita*, based on Eurostat 2007 pop. estimates (Luxembourg not shown).

Net contributors

-5000 to -1000 euro per capita
-1000 to -500 euro per capita
-500 to 0 euro per capita

Net recipients

0 to 500 euro per capita
500 to 1000 euro per capita
1000 to 5000 euro per capita
5000 to 10000 euro per capita
10000 euro plus per capita

One of the first cartographers to generate cartograms with the aid of computer visualization was Waldo Tobler of UC Santa Barbara in the 1960s. Prior to Tobler's work, cartograms were created by hand (as they occasionally still are). The National Center for Geographic Information and Analysis located on the UCSB campus maintains an online Cartogram Central with resources regarding cartograms.

A number of software packages generate cartograms. Most of the available cartogram generation tools work in conjunction with other GIS software tools as add-ons or independently produce cartographic outputs from GIS data formatted to work with commonly used GIS products. Examples of cartogram software include ScapeToad, Cart, and the Cartogram Processing Tool (an ArcScript for ESRI's ArcGIS), which all use the Gastner-Newman algorithm. An alternative algorithm, Carto3F, is also implemented as an independent program for non-commercial use on Windows platforms. This program also provides an optimization to the original Dougenik rubber-sheet algorithm. The CRAN package recmap provides an implementation of a rectangular cartogram algorithm.

Cartograms can also be constructed manually, either by hand or in a computer-assisted environment. Block cartograms are constructed by arranging geometrically regular equal-sized blocks, with the number of blocks allocated to each district proportional to the population variable. Several examples of block cartograms were published during the 2016 U.S. presidential election season by *The Washington Post* , the *FiveThirtyEight* blog , and the *Wall Street Journal* , among others.

Geologic Map

Geologic maps are not like other maps. Geologic maps, like all maps, are designed to show where things are. But, whereas the maps we know best show the distribution of roads or rivers or county boundaries, a geologic map shows the distribution of geologic features, including different kinds of rocks and faults. A geologic map is usually printed on top of a regular map (called a base map) to help you locate yourself on the map. The base map is printed with light colors, so it doesn't interfere with seeing the geologic features on the map. The geology is represented by colors, lines, and special symbols unique to geologic maps. Understanding these features will allow you to understand much of the geology shown in almost any standard geologic map.

They are used in land-use planning, geohazards, geohistory, water/energy/mineral resource development, waste disposal, and national defense, which in turn are used to solve the many problems and challenges that affect our quality of life. Improved and more detailed geologic maps provide more accurate data, leading to more reliable conclusions. Geologic maps are fundamental elements for informing the policy decisions of federal, state, and local agencies. Positive benefit-to-cost ratios as much as 50:1 have been determined for geologic maps.

Although geologic maps are typically depicted in two dimensions, a third dimension is represented with topographic contours, cross sections, block diagrams, and virtual reality displays (rotating transparent block diagrams). The fourth dimension of time can also be shown through sequential cross sections, chronostratigraphic diagrams, and displays that go by names like kinematic and retrodeformation. Maps can be tailored to fit the needs of specific projects such as bedrock maps, surficial deposit maps, or material engineering properties maps.

Today most geologic maps are produced in digital forms such as geographic information system (GIS) databases, downloadable Internet files, and interactive website maps that provide users with many options, increasing availability and usefulness. For example, in GIS format one might sum the areas of landslides in a geologic unit and divide by the area of the geologic unit to get a relative idea of which units are most prone to landslides.

Colored Areas

The most striking features of geologic maps are its colors. Each color represents a different geologic unit. A geologic unit is a volume of a certain kind of rock of a given age range. So a sandstone of one age might be colored bright orange, while a sandstone of a different age might be colored pale brown. Many geologic units are given names that relate to where their characteristics are best displayed, or where they were first studied. For example, the Briones sandstone was first described in Briones Valley,California.

Some geologic units have not yet been named, so those are identified with terms related to the kind of rock in the unit like 'Sandstone and shale,' 'Unnamed sandstone', or 'Undivided shale'. But all units, named and unnamed, have a color on the geologic map, and the area of a given color is the area where that geologic unit is the one at the surface (usually the soil on top of the rocks is disregarded).

Geologic units are named and defined by the geologists who made the geologic map, based on their observations of the kinds of rocks and their investigations of the age of the rocks. As more information is gathered, perhaps by other geologists, new geologic units might be defined. These disagreements can be a basis for scientific progress, and illustrate the need for continuing to investigate the geology of an area.

Letter Symbols

In addition to color, each geologic unit is assigned a set of letters to symbolize it on the map. Usually the symbol is the combination of an initial capital letter followed by one or more small letters. The capital letter represents the age of the geologic unit. Geologists have divided the history of the Earth into Eons (the largest division), Eras, Periods, and Epochs , mostly based on the fossils found in rocks. The most common division of time used in letter symbols on geologic maps is the Period. Rocks of the four most recent Periods are found in the San Francisco Bay area shown on this map, so most letter symbols begin with a capital letter representing one of the four Periods: J (Jurassic - 195 to 141 million years ago), K (Cretaceous - 141 to 65 million years ago), T (Tertiary - 65 to 2 million years ago), or Q (Quaternary - 2 million years ago until today).

Occasionally the age of a rock unit will span more than one period, if the period of many years required to create a body of rock happens to fall on both sides of a time boundary. In that case both capital letters are used. For example, QT would indicate that the rock unit began to form in Tertiary time and was completed in Quaternary time. The few geologic units formed an unknown amount of time ago have letter symbols with no capital letters.

The small letters indicate either the name of the unit, if it has one, or the type of rock, if the unit has no name. So Kjm would be the symbol for the Joaquin Miller sandstone (formed in the Cretaceous Period), while Ks would be the symbol for an unnamed unit of shale formed in the same Period, and gb would be the symbol for gabbro (a dark-colored igneous rock) of unknown age.

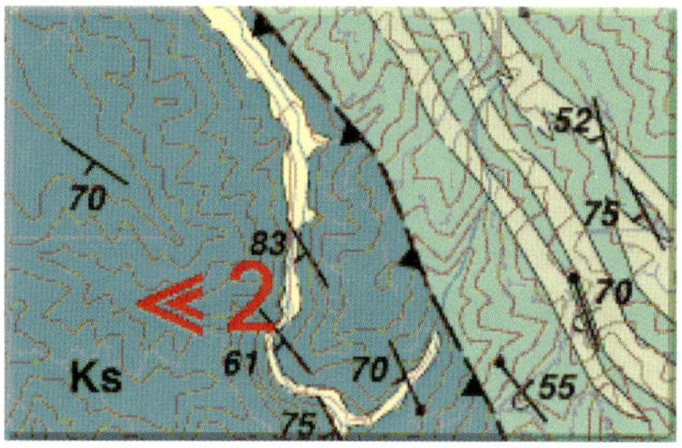

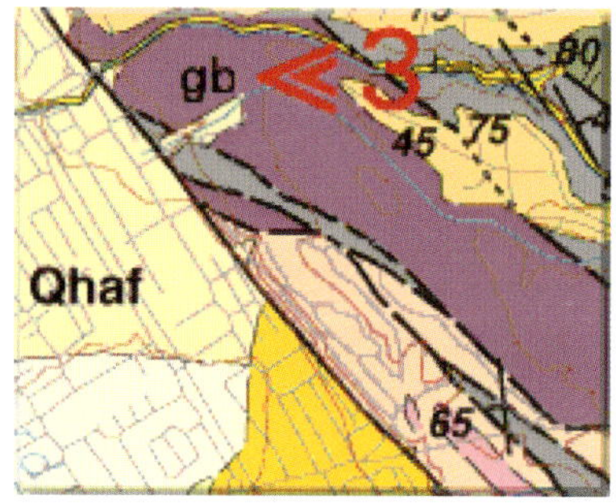

Lines on the Map

- Contact Lines

The place where two different geologic units are found next to each other is called a contact, and that is represented by different kinds of lines on the geologic map. The two main types of contacts shown on most geologic maps are depositional contacts and faults.

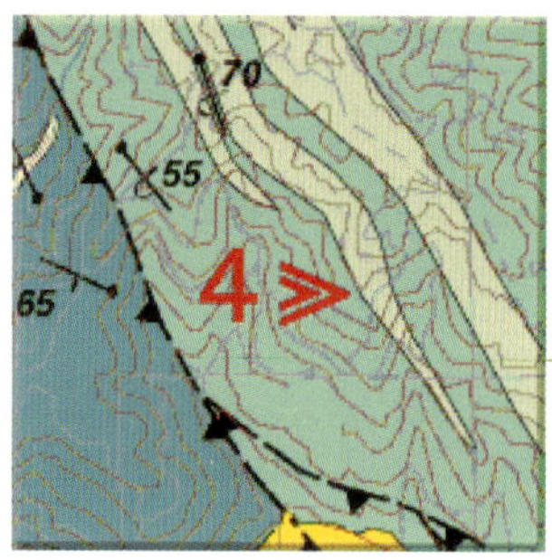

All geologic units are formed over, under, or beside other geologic units. For example, lava from a volcano flows over the landscape, and when the lava hardens into rock, the place where the lava-rock rests on the rocks underneath is a depositional contact. Where the original depositional contact between geologic units is preserved, it is shown on the geologic map as a thin line (location 4).

- Faults

However, in geologically active areas like the San Francisco Bay area, geologic units tend to be broken up and moved along faults (it is fault movements that cause earthquakes!). When different geologic units have been moved next to one another after they were formed, the contact is a fault contact, which is shown on the map by a thick line (location 5). Faults can cut through a single geologic unit. These faults are shown with the same thick line on the map, but have the same geologic unit on both sides.

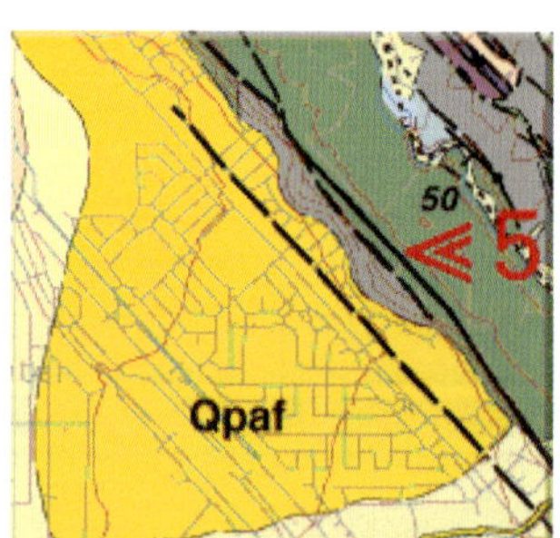

Remember, just because the map shows a fault doesn't mean that fault is still active and is likely to cause an earthquake. Rocks can preserve records of faults that have been inactive for many millions of years. But knowing where the faults are is the first step toward finding the ones that can move. Special geologic maps of the faults known to be still moving are constantly being upgraded here at the United States Geological Survey, as well as by State geological surveys and university researchers.

Other Lines

- Folds

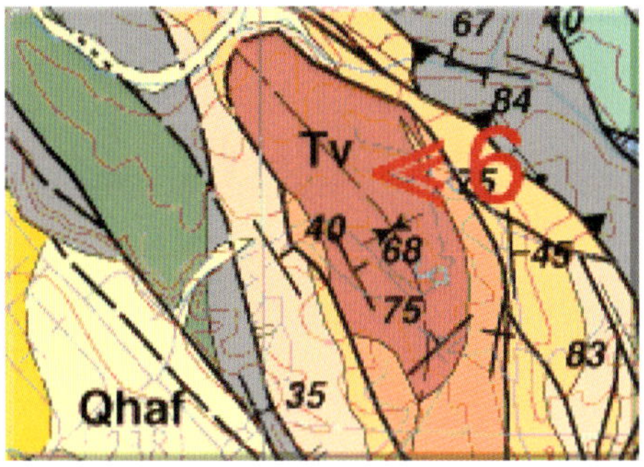

Another kind of line shown on most geologic maps is a fold axis. In addition to being moved by faults, geologic units can also be bent and warped by the same forces into rounded wavelike shapes called folds. A line that follows the crest or trough of the fold is called the fold axis. This is marked on a geologic map with a line a little thicker than a depositional contact, but thinner than a fault (location 6).

- Solid, dashed, or dotted lines

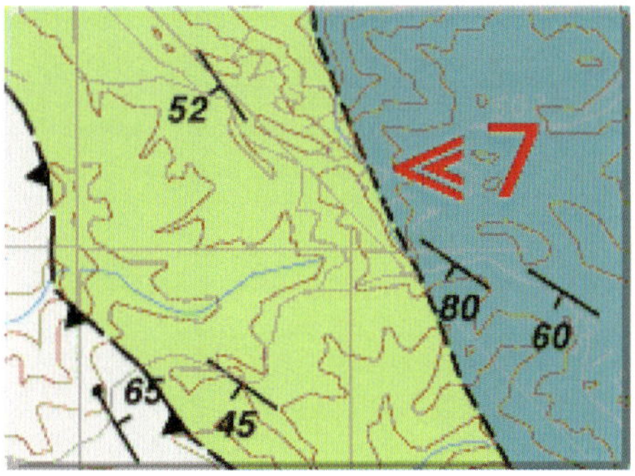

All thicknesses of lines are also modified by being solid, dashed, or dotted. Often contacts are obscured by soil, vegetation, or human construction. Those places where the line is precisely located it is shown as solid, but where it is uncertain it is dashed (location 7). The shorter the dash, the more uncertain the location. A dotted line is the most uncertain of all, because it is covered by a geologic unit, so no amount of searching at the surface could ever locate it (location 8). The lines on the map may also be modified by other symbols on the line (triangles, small tic marks, arrows, and more) which give more information about the line. For example, faults with triangles on them (location 9) show that the side with the triangles has been thrust up and over the side without the triangles (that kind of fault is called a reverse fault or a thrust fault). All the different symbols on the lines are explained in the map key.

Strike and Dip

Many kinds of rocks form in broad, flat layers, called beds, that stack up like the layers of a cake. In areas like northern Arizona, thick stacks of rock beds that have built up over millions of years remain in their original flat orientation (where they can be viewed as multicolored horizontal layers of rock that make up the spectacular walls of the Grand Canyon). In places like California near active plate boundaries, however, the forces that make earthquakes don't leave the beds flat for long, but bend and tilt them.

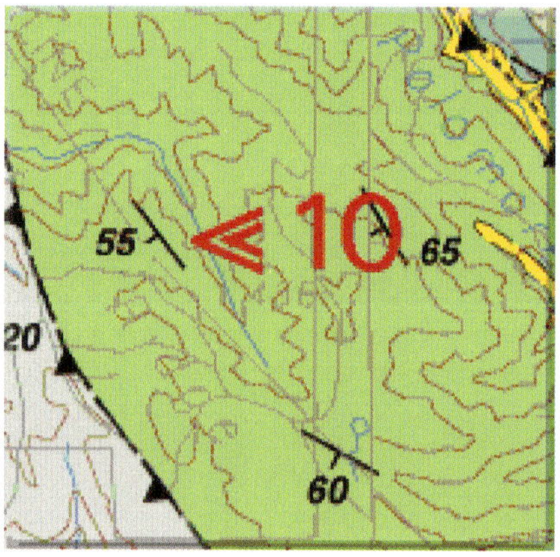

Tilted beds are shown on a geological map with a strike and dip symbol (location 10). The symbol consists of three parts: a long line, a short line, and a number. The long line is called the strike line, and shows the direction in the bed that is still horizontal. Any tilted surface has a direction that is horizontal (think about walking on the side of a hill, there is always a way to go that is neither up nor down, but is level).

The strike line shows that horizontal direction in the beds. The short line is called the dip line, and shows which way the bed is tilted. The number is called the dip, and shows how much the bed is tilted, in degrees, from flat. The higher the number, the steeper the tilting of the bed, all the way up to 90 degrees if the bed is tilted all the way onto its side. Strike and dip symbols can be modified to give more information about the tilted beds just like lines can be, and these modifications are also explained in the MAP KEY.

Map Key

All geologic maps come with a table called a map key. In the map key, all the colors and symbols are shown and explained. The map key usually starts with a list showing the color and letter symbol of every geologic unit, starting with the youngest or most recently formed units (in the example map those are the man-made deposits), along with the name of the unit (if it has one) and a short description of the kinds of rocks in that unit and their age (in the key, the age is described by Epochs, subdivisions of the Periods shown in the letter symbol). After the list of geologic units, all the different types of lines on the map are explained, and then all the different strike and dip symbols. The map key will also include explanations of any other kinds of geologic symbols used on a map (locations where fossils were found, locations of deposits of precious metals, location of faults known to be active, and any other geologic feature that might be important in the area shown by the geologic map). Because the geology in every area is different, the map key is vital to understanding the geologic map.

Symbols

Lithologies

Rock units are typically represented by colors. Instead of (or in addition to) colors, certain symbols can be used. Different geologic mapping agencies and authorities have different standards for the colors and symbols to be used for rocks of differing types and ages.

Orientations

A standard Brunton Geological compass, used commonly by geologists

Geologists take two major types of orientation measurements (using a hand compass like a Brunton compass): orientations of planes and orientations of lines. Orientations of planes are often measured as a "strike" and "dip", while orientations of lines are often measured as a "trend" and "plunge".

Strike and dip symbols consist of a long "strike" line, which is perpendicular to the direction of greatest slope along the surface of the bed, and a shorter "dip" line on side of the strike line where the bed is going downwards. The angle that the bed makes with the horizontal, along the dip direction, is written next to the dip line. In the azimuthal system, strike and dip are often given as "strike/dip" (for example: 270/15, for a strike of west and a dip of 15 degrees below the horizontal).

Trend and plunge are used for linear features, and their symbol is a single arrow on the map. The arrow is oriented in the downgoing direction of the linear feature (the "trend") and at the end of the arrow, the number of degrees that the feature lies below the horizontal (the "plunge") is noted. Trend and plunge are often notated as PLUNGE → TREND (for example: 34 → 86 indicates a feature that is angled at 34 degrees below the horizontal at an angle that is just East of true South).

Maps and Mapping Around the Globe

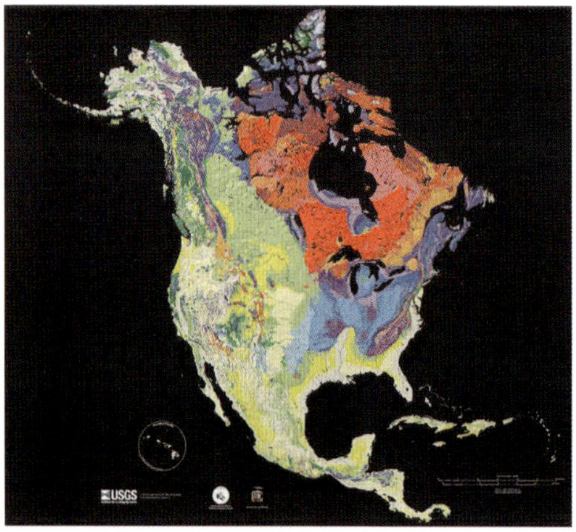

Geologic map of North America superimposed on a shaded relief map

United States

In the United States, geologic maps are usually superimposed over a topographic map (and at times over other base maps) with the addition of a color mask with letter symbols to represent the kind of geologic unit. The color mask denotes the exposure of the immediate bedrock, even if obscured by soil or other cover. Each area of color denotes a geologic unit or particular rock formation (as more information is gathered new geologic units may be defined). However, in areas where the bedrock is overlain by a significantly thick unconsolidated burden of till, terrace sediments, loess deposits, or other important feature, these are shown instead. Stratigraphic contour lines, fault lines, strike and dip symbols, are represented with various symbols as indicated by the map key. Whereas topographic maps are produced by the United States Geological Survey in conjunction with the states, geologic maps are usually produced by the individual states. There are almost no geologic map resources for some states, while a few states, such as Kentucky and Georgia, are extensively mapped geologically. Technically A map that uses colors.

United Kingdom

In the United Kingdom the term *geological map* is used. The UK and Isle of Man have been extensively mapped by the British Geological Survey (BGS) since 1835; a separate Geological Survey of Northern Ireland (drawing on BGS staff) has operated since 1947.

Two 1:625,000 scale maps cover the basic geology for the UK. More detailed sheets are available at scales of 1:250,000, 1:50,000 and 1:10,000. The 1:625,000 and 1:250,000 scales show both onshore and offshore geology (the 1:250,000 series covers the entire UK continental shelf), whilst other scales generally cover exposures on land only.

Sheets of all scales (though not for all areas) fall into two categories:

1. Superficial deposit maps (previously known as *solid and drift* maps) show both bedrock *and* the deposits on top of it.

2. Bedrock maps (previously known as *solid* maps) show the underlying rock, without superficial deposits.

The maps are superimposed over a topographic map base produced by Ordnance Survey (OS), and use symbols to represent fault lines, strike and dip or geological units, boreholes etc. Colors are used to represent different geological units. Explanatory booklets (memoirs) are produced for many sheets at the 1:50,000 scale.

Small scale thematic maps (1:1,000,000 to 1:100,000) are also produced covering geochemistry, gravity anomaly, magnetic anomaly, groundwater, etc.

Although BGS maps show the British national grid reference system and employ an OS base map, sheet boundaries are not based on the grid. The 1:50,000 sheets originate from earlier 'one inch to the mile' (1:63,360) coverage utilising the pre-grid Ordnance Survey One Inch Third Edition as the base map. Current sheets are a mixture of modern field mapping at 1:10,000 redrawn at the 1:50,000 scale and older 1:63,360 maps reproduced on a modern base map at 1:50,000. In both cases the original OS Third Edition sheet margins and numbers are retained. The 1:250,000 sheets are defined using lines of latitude and longitude, each extending 1° north-south and 2° east-west.

Singapore

The first geological map of Singapore was produced in 1974, produced by the then Public Work Department. The publication includes a locality map, 8 map sheets detailing the topography and geological units, and a sheet containing cross sections of the island.

Since 1974, for 30 years, there were many findings reported in various technical conferences on new found geology islandwide, but no new publication was produced. In 2006, Defence Science & Technology Agency, with their developments in underground space promptly started a re-publication of the Geology of Singapore, second edition. The new edition that was published in 2009, contains a 1:75,000 geology map of the island, 6 maps (1:25,000) containing topography, street directory and geology, a sheet of cross section and a locality map.

The difference found between the 1976 Geology of Singapore report include numerous formations found in literature between 1976 and 2009. These include the Fort Canning Boulder Beds and stretches of limestone.

Topographic Map

Topographic maps are detailed, accurate graphic representations of features that appear on the Earth's surface. These features include:

- cultural: roads, buildings, urban development, railways, airports, names of places and geographic features, administrative boundaries, state and international borders, reserves
- hydrography: lakes, rivers, streams, swamps, coastal flats

- relief: mountains, valleys, contours and cliffs, depressions
- vegetation: wooded and cleared areas, vineyards and orchards.

A map legend (or key) lists the features shown on that map, and their corresponding symbols.

Topographic maps usually show a geographic graticule and a coordinate grid, so you can determine relative and absolute positions of mapped features.

It is important to note that a map is merely a two or three dimensional representation of the physical environment at a given time. Therefore, a map will never be entirely up to date. Changes to the landscape and cultural features regularly occur (such as roads, vegetation, and buildings), resulting in maps becoming dated, although the rate of obsolescence varies depending upon location.

There are three major factors that aid in the visualization of terrain from a 2d map:

- Contour Lines
- Colors
- Shading

The ability to fuse these major factors is the most critical skill one can learn when using a topographic map. The primary purpose of a topographic map is to accurately represent the shape of the Earth's surface, but the utility doesn't stop there. Topographic maps also represent streets and trails, vegetation, streams, and every type of feature that may positively or negatively impact your ability to navigate through the terrain.

Contour Lines

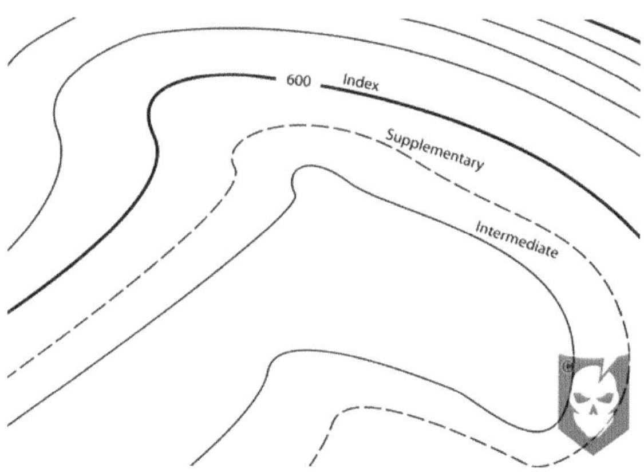

Contour lines are imaginary; they are map artifacts used to represent paths or segments of Earth at an equal elevation. These paths and segments are presented as elevations (vertical distance above or below sea level) and reliefs (the shape of terrain features on the Earth's surface).

Not all contour lines are created equal. Heavier contour lines are known as *indexed* contour lines and are normally numbered showing elevation. Typically every fifth contour line is an index.

Lighter contour lines, that fall between indexed lines, are known as *intermediate* contour lines. These lines do not have their elevation given and are found in sets of four between indexed contour lines.

Finally, when the terrain is expansively flat, cartographers will often include *supplementary* contour lines, which are dashed lines indicating an elevation that is half of the elevation between the contour lines surrounding it. They are typically found where there is little change in elevation.

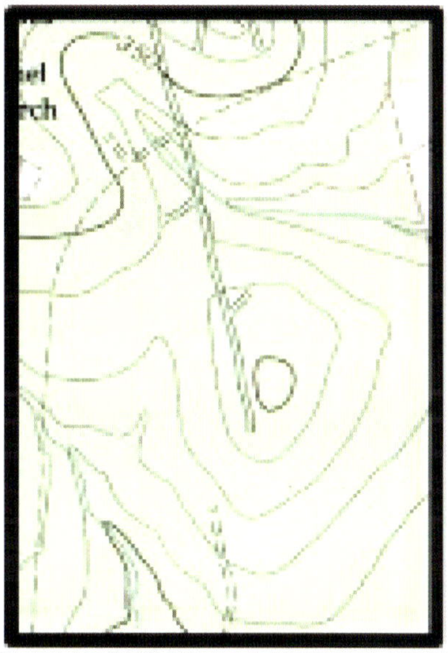

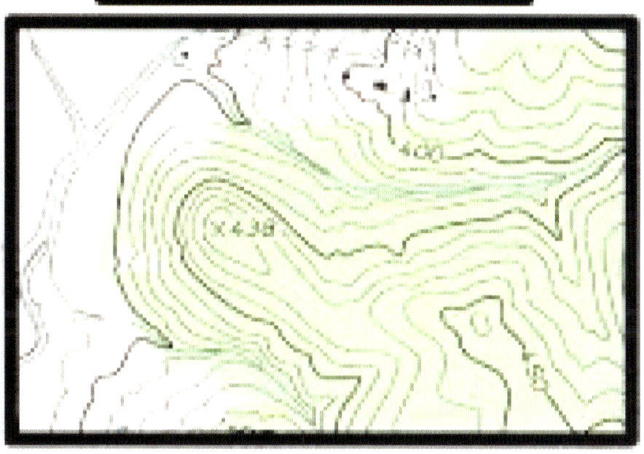

The key thing to remember about contour lines is this: *The closer the spacing, the more rapid the change in elevation.* For a leisurely Sunday stroll, you might be looking to cross just a few contour lines, or perhaps follow a trail that shadows a few contour lines on the map. For rock climbing, look for a concentration of contour lines in a single area. If you're looking for a true vertical cliff, look for a series of contour lines drawn so closely together that they appear to be a single line.

The Huntsville quad doesn't really have any hard elevation terrain (though navigating swamps can be pretty hard). Notice the foot trail along the top image above. It intersects two indexed contour lines, but the distance between the indexed contours is pretty significant; it's easy to see that this is a very easy path to navigate with respect to elevation gain.

In contrast, the bottom graphic shows a hill where the top is at 438'. The indexed contour line to the immediate left is 400', and the one further left is 350'. Walking up the western face of this hill would be more challenging than traversing the easy foot trail.

Topographic Map Colors

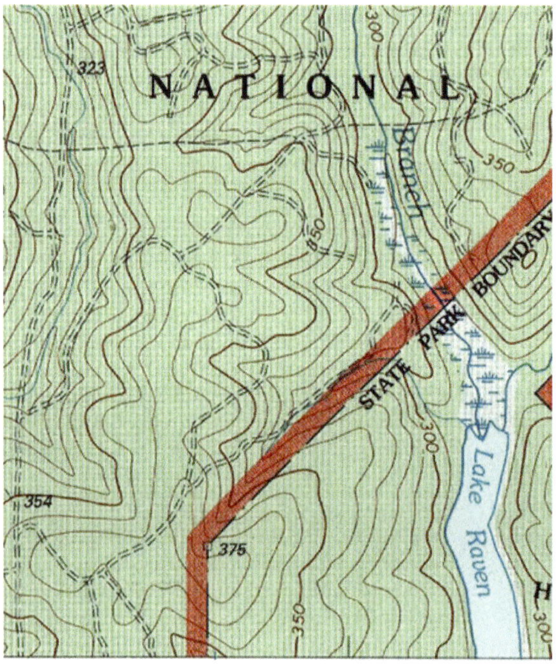

It's important to know what kind of terrain and environment you're traveling into and what the map of that area is telling you.

The color brown is used to denote most contour lines on a map, which are relief features and elevations. Topographic maps use green to denote vegetation such as woods, while blue is used to denote water features like lakes, swamps, rivers, and drainage.

At higher elevations, mountains may be snow-capped year around, or the terrain may actually be a glacier. In each of these cases, contour lines are also drawn in blue. It is therefore possible to quickly discern that a particular route from A to B might be more treacherous than operating at a high altitude—the trek might require crampons, an ice axe, and other materials that might not be readily available once in the backcountry.

Finally, black is used to represent man-made objects, including trails. Red is used for man-made features, like main roads or political boundaries, and purple for new changes or updates on the map that weren't previously represented. Newer maps no longer use purple, but since so many older maps exist, it's worth mentioning.

At the bottom of the Huntsville quad is Lake Ravenshaded in blue, since it's a water feature. Notice the thick red line, delineating a political feature, the state park boundary. The majority of the image is shaded green, showing that this part of the map is full of vegetation. Contour lines are present in brown, as are two different types of paths: an unimproved or 4-wheel-drive trail represented by parallel dashed lines and a foot trail represented with a single dashed line.

Admittedly outside of our topographic scope, it's worth noting that the USGS also provides extensive documentation on colors (and shading) for representing geologic features. This could be particularly useful for those interested in either rock climbing or geology in general

Shading

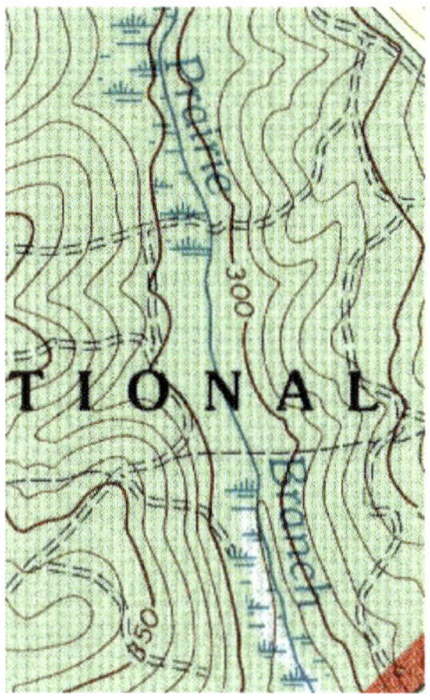

Color similarity between features does not mean that the features are equivalent. Due north of Lake Raven is the Prairie Branch, another name for a stream. Other names that equate to a stream include kill, run, fork, and brook. What's interesting about Prairie Branch is that it has led to the formation of a wooded marsh or swamp.

Navigating across Prairie Branch could be difficult. Since this is Texas, expect to run into water moccasins, copperheads, and perhaps the occasional alligator, among all of the other friendly animals that call Sam Houston National Forest home.

Terrain Association

Orienting the map is an exercise that rotates the map so that north on the map is aligned with north in the real world. Orienting a map is critical because it allows you to point in a direction and know with confidence what terrain lies ahead. But if your compass breaks, how will you know where you're headed?

It happens! You may find yourself in the backcountry with a map, but that fall (more like a slide) down that last rock face shattered your compass. Now what? Well, the key is to read the terrain and orient your map, a skill called *terrain association*. This task is far easier in mountainous or hilly areas than in areas where there is little to no reference, like in the plains or in a rain forest where your view is blocked.

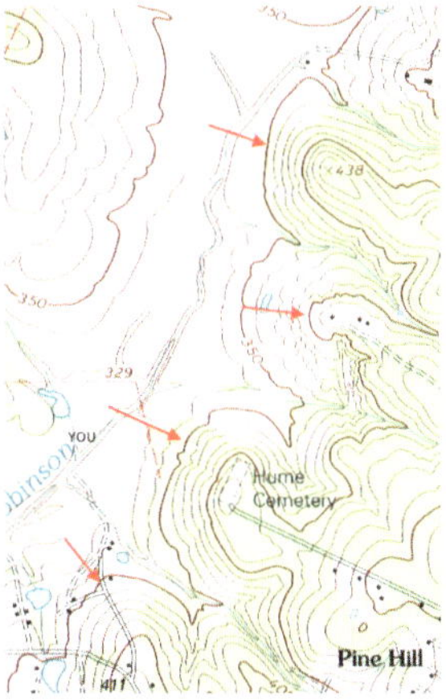

The Huntsville quad isn't the Rocky Mountains, so orienting a map is more challenging because you can't simply look around and pick out the tallest peaks. That doesn't mean that orienting the map isn't possible. Notice how the map helps you visually see a valley? Close your eyes and imagine standing in the flat valley, somewhere near the 'n' in Robinson, then look east.

The map tells us that the valley isn't heavily vegetated because it isn't in green, but the hills to the east are. We also can tell from the contour lines that there is a significant elevation change of about 100'. It then becomes possible to visualize a slinking, rapidly rising set of hills with at least four distinct faces. Valleys are often easy to identify because there is typically a water feature running through the middle; the water feature is typically responsible for carving out the valley.

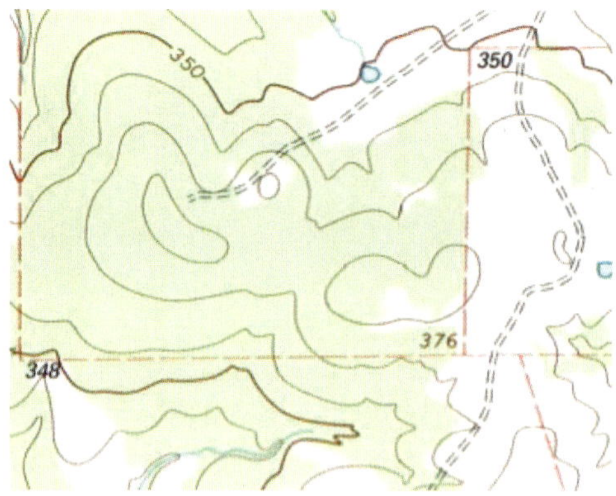

In the above image we can see two hills sitting across from each other. The change in elevation isn't as dramatic as one might find in Utah or Colorado, but it's a saddle. The key to identifying a saddle is to look for concentric circles with a space between them. Someone once told me that a saddle can be thought of as a frying pan with two eggs on it. There's the surface of the pan, then there's an elevation gain up to the egg whites and then there's the top of the two hills, the yolks. It's a silly but effective way to understand that the saddle is the dip between the two areas of higher terrain.

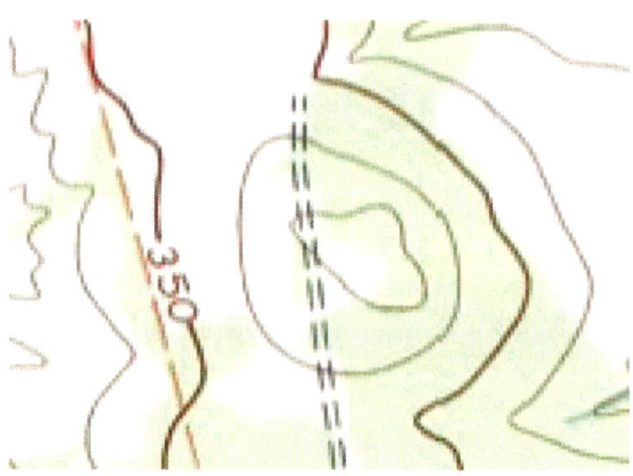

Hills also stand out on a map and are shown as single concentric circles. Above is an example of a very small hill with only about 20' above its surrounding terrain.

Reading terrain isn't difficult, but it requires the ability to close your eyes and envision the surrounding area sometimes. Other terrain features can be picked out on a topographic map, including cliffs, spurs, depressions, ridge lines, and draws.

Uses

Curvimeter used to measure the length of a curve

Topographic maps have multiple uses in the present day: any type of geographic planning or large-scale architecture; earth sciences and many other geographic disciplines; mining and other earth-based endeavours; civil engineering and recreational uses such as hiking and orienteering.

Conventions

The various features shown on the map are represented by conventional signs or symbols. For example, colors can be used to indicate a classification of roads. These signs are usually explained in the margin of the map, or on a separately published characteristic sheet.

Topographic maps are also commonly called *contour maps* or *topo maps*. In the United States, where the primary national series is organized by a strict 7.5-minute grid, they are often called *topo quads* or quadrangles.

Topographic maps conventionally show topography, or land contours, by means of contour lines. Contour lines are curves that connect contiguous points of the same altitude (isohypse). In other words, every point on the marked line of 100 m elevation is 100 m above mean sea level.

These maps usually show not only the contours, but also any significant streams or other bodies of water, forest cover, built-up areas or individual buildings (depending on scale), and other features and points of interest.

Today, topographic maps are prepared using photogrammetric interpretation of aerial photography, lidar and other Remote sensing techniques. Older topographic maps were prepared using traditional surveying instruments.

Dymaxion Map

Humans have always had a particular fascination with understanding the world around them, and this appeal can be seen in the large number of maps invented throughout history starting with the ancient Greeks. All different varieties of maps have appeared, highlighting certain geographical, cultural, political, or social phenomena. Maps are also a common part of our everyday lives. Maps, however, in themselves have a fundamental flaw: maps are almost always incomplete representations of the earth since the earth is a sphere and maps are usually two-dimensional flat surfaces.

Over the years, there are been many attempts to create a map projection that more fully represents the Earth in its relative shape, distance, area, and direction. The Mercator world map and Robinson projection are popular map projections

However, these maps still have some extreme distortions. For example, in the Mercator projection, the poles are problematic with Greenland appearing to be three times its relative size and Antarctica appearing as a long, thin, white strip across the bottom. In the Robinson projection, Greenland still appears to be larger than its actual size. Distortions continue to be a problem for every two-dimensional map.

What is a Dymaxion Map?

One attempt to make a better map with fewer distortions is called the Dymaxion map, also known as the Fuller Map Projection. This map was created by Buckminster Fuller, an American architect, author, and inventor. What makes the map unique is that the Earth is projected onto the surface of an icosahedron, a polyhedron that is comprised of twenty triangular faces and thirty edges. This unique shape can be flattened into two dimensions but also folded to create a three dimensional object. When flattened, the Dymaxion map shows the Earth's continents as almost one contiguous land mass or island in the middle of one ocean. When folded, the map is a rough three dimensional representation of the earth.

The Dymaxion map from Buckminster Fuller stands out because it is the only flat map of the Earth's surface in its entirety that does not contain any obvious visual distortions of the shapes and sizes of land masses relative to their global scales. It is also does this without splitting any of the continents. The map, however, does have an unusual configuration in order to preserve shapes and sizes. The Dymaxion map was only intended to be used for representations of the complete globe. Each triangle edge of the Fuller Projection Map lines up with the scale of a partial great circle on a corresponding globe, and the other points within each of its facets shrink towards the middle instead of enlarging around the edges like other map projections.

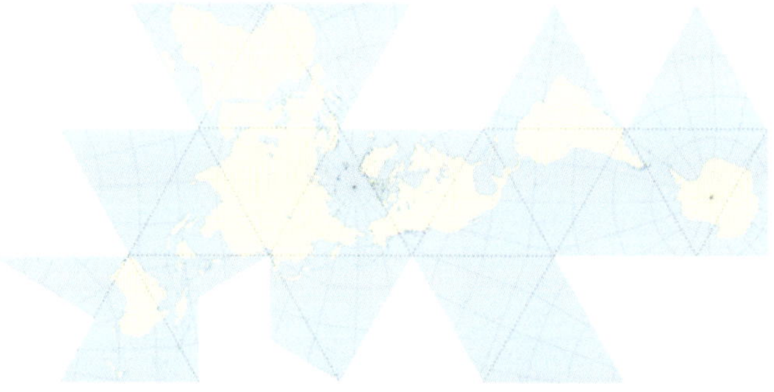

Unfolded Dymaxion Map (dashed lines indicate fold points) by Eric Gaba

Buckminster Fuller worked on the map for several decades before applying and receiving a patent for the Dymaxion map in 1946. The original projection was onto a cuboctahedron, but then in 1954, Fuller used a modified icosahedron which is still in use today.

Fuller completed the map after working on it for several decades. He appeared to understand the fundamental problem with maps and wanted to create a map that would not grossly distort the relative size and shapes of the different continents. According to Fuller, the primary function of a Dymaxion map was to allow people to view the land masses without dividing them up. He also wanted to create a map that can be unfolded in many different ways in order to emphasize many different aspects of the world. Plus, the way the Earth appears as one large land mass surrounded when the map is flattened depicts how ancient peoples thought about the world. Fuller wanted to create the best visual representation of what he called Spaceship Earth.

Maps are also problematic in that they can reflect a cultural or political bias. Traditional maps have been criticized for reinforcing elements that separate humanity and supporting us versus them thinking, self-interests, and isolation. These traditional maps often fail to represent the patterns of the increasing globalization and growing relationships across continents. Fuller himself was critical of the way maps exhibited the northern hemisphere as superior and the southern hemisphere as inferior. He argued that his Dymaxion map had no right way up since there is no up or down in the universe.

Properties

Fuller claimed that his map had several advantages over other projections for world maps.

It has less distortion of relative size of areas, most notably when compared to the Mercator projection; and less distortion of shapes of areas, notably when compared to the Gall–Peters projection. Other compromise projections attempt a similar trade-off.

More unusually, the Dymaxion map does not have any "right way up". Fuller argued that in the universe there is no "up" and "down", or "north" and "south": only "in" and "out". Gravitational forces of the stars and planets created "in", meaning "towards the gravitational center", and "out", meaning "away from the gravitational center". He attributed the north-up-superior/south-down-inferior presentation of most other world maps to cultural bias.

Fuller intended the map to be unfolded in different ways to emphasize different aspects of the world. Peeling the triangular faces of the icosahedron apart in one way results in an icosahedral net that shows an almost contiguous land mass comprising all of Earth's continents – not groups of continents divided by oceans. Peeling the solid apart in a different way presents a view of the world dominated by connected oceans surrounded by land.

Showing the continents as "one island earth" also helped Fuller explain, in his book *Critical Path*, the journeys of early seafaring people, who were in effect using prevailing winds to circumnavigate this world island.

However, the Dymaxion map can also prove difficult to use. It is, for example, confusing to describe the four cardinal directions and locate geographic coordinates. The awkward shape of the map may be counterintuitive to most people trying to use it. For example, tracing a path from India to Chile may be confusing. Depending on how the map is projected, land masses and oceans are often divided into several pieces.

Electronic Maps

The digital map is an electronic map, which operation is based on a combination of graphic elements assigned to it in the form of electronic information. It is based on naturally harvested and processed into digital cartographic data. Advantages of digital maps compared with paper maps are visible at first glance:

- digital maps can include any area,
- they're scalable - we can freely zoom them in and out,
- they're more accurate (there are no printed maps on such a scale as the greatest closer of digital maps),
- they are more up-to-date (can be remotely updated at a much lower cost, and labour),
- they may contain a lot of different information, and layered architecture that allows you to group data and turning them on and off doesn't disrupt their visibility,
- they are interactive (answer to the selected user actions - eg. Click),
- they take up a lot less spaceJ and we may be always with you.

The creation of high-quality digital map requires not only cartographic skills and resources, but also the know-how and experience in the field of geoinformatics.

The Two Types of Electronic Map

Raster map: Simply an electronic picture of the equivalent paper map. It contains no 'intelligence', is not 'navigable' and can only be printed or viewed like a paper map.

Vector (or navigable) map: a structured database containing all sorts of data, including road geometry (the road's physical shape); road attributes (street name, traffic flow direction, turn restrictions, etc); city names; points of interest (POI's); postcodes, house numbers, etc.

Creating an electronic map

Vector map creation is a long and complex process that combines desk-based map processing and in-the-field surveying:

1. The paper or raster map is first digitised by a data-processing team. This involves 'tracing' all the roads.
2. All points on the map are then geo-coded (i.e. converted into longitude and latitude).
3. The data base is then 'attributed' by adding details of each road: its street name,traffic flow directions, etc.

More attributes are added from desk-based research into governmental and other data.

4. Finally the roads are surveyed. This involves someone actually driving each road and recording - usuaWlly by differential GPS which pinpoints locations to within 50 centimetres and video camera - the road restrictions, traffic flow, signs, speed limits, turn offs, etc.

Data Collection

Digital maps heavily rely upon a vast amount of data collected over time. Most of the information that comprise digital maps is the culmination of satellite imagery as well as street level information. Maps must be updated frequently to provide users with the most accurate reflection of a location. While there is a wide spectrum on companies that specialize in digital mapping, the basic premise is that digital maps will accurately portray roads as they actually appear to give "life-like experiences."

Functionality and Use

Computer Applications

Proprietary and non-proprietary computer programs and applications provide imagery and street-level map data for much of the world.

Scientific Applications

The development of mobile computing (PDAs, tablet PCs, laptops, etc.) has recently (since about 2000) spurred the use of digital mapping in the sciences and applied sciences. As of 2009, science fields that use digital mapping technology include geology, engineering, architecture, land surveying, mining, forestry, environmental, and archaeology.

GPS Navigation Systems

The principal use by which digital mapping has grown in the past decade has been its connection to Global Positioning System (GPS) technology. GPS is the foundation behind digital mapping navigation systems.

How it Works

The coordinates and position as well as atomic time obtained by a terrestrial GPS receiver from GPS satellites orbiting Earth interact together to provide the digital mapping programming with points of origin in addition to the destination points needed to calculate distance. This information is then analyzed and compiled to create a map that provides the easiest and most efficient way to reach a destination.

More technically speaking, the device operates in the following manner:

1. GPS receivers collect data from at least four GPS satellites orbiting the Earth, calculating position in three dimensions.
2. The GPS receiver then utilizes position to provide GPS coordinates, or exact points of latitudinal and longitudinal direction from GPS satellites.
3. The points, or coordinates, output an accurate range between approximately "10-20 meters" of the actual location.
4. The beginning point, entered via GPS coordinates, and the ending point, (address or coordinates) input by the user, are then entered into the digital mapping software.
5. The mapping software outputs a real-time visual representation of the route. The map then moves along the path of the driver.
6. If the driver drifts from the designated route, the navigation system will use the current coordinates to recalculate a route to the destination location.

Map Projection

A *map projection* is a mathematically described technique of how to represent the Earth's curved surface on a flat map. To represent parts of the surface of the Earth on a flat paper map or on a computer screen, the curved horizontal reference surface must be mapped onto the 2D mapping plane. The reference surface for large-scale mapping is usually an oblate ellipsoid, and for small-scale mapping, a sphere. Mapping onto a 2D mapping plane means transforming each point on the reference surface with geographic coordinates (φ,λ) to a set of Cartesian coordinates (x,y) representing positions on the map plane (figure below).

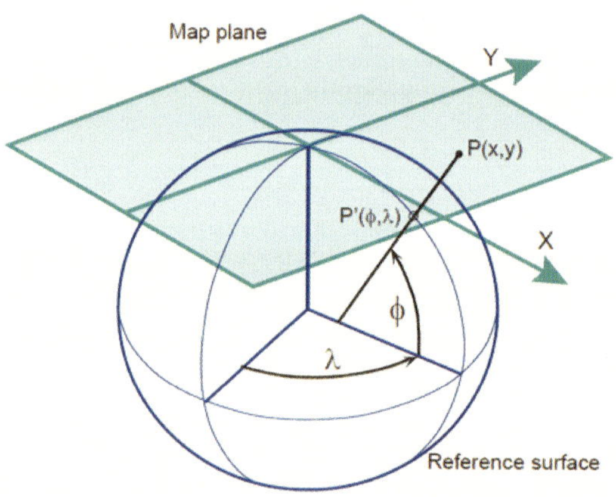

Example of a map projection where the reference surface with geographic coordinates (φ,λ) is projected onto the 2D mapping plane with 2D Cartesian coordinates (x, y).

The actual mapping cannot usually be visualized as a true geometric projection, directly onto the mapping plane as illustrated in the figure above. This is mostly achieved through mapping equations. A *forward mapping equation* transforms the geographic coordinates (φ,λ) of a point on the curved reference surface to a set of planar Cartesian coordinates (x,y), representing the position of the same point on the map plane:

$$(x, y) = f(\varphi, \lambda)$$

The corresponding *inverse mapping equation* transforms mathematically the planar Cartesian coordinates (x,y) of a point on the map plane to a set of geographic coordinates (φ,λ) on the curved reference surface:

$$(\varphi, \lambda) = f(x, y)$$

Following are two examples of mapping equations for the sphere (equations for the ellipsoid are generally more complex).

The first example are the mapping equations used for the *Mercator* projection: The forward mapping equation is:

$$x = R(\lambda - \lambda_0)$$

$$y = R\left(ln\left(tan\left(\frac{\pi}{4}+\frac{\phi}{2}\right)\right)\right)$$

The inverse mapping equation is:

$$\phi = \frac{\pi}{2} - 2\arctan\left(e^{\frac{-y}{R}}\right)$$

$$\lambda = \frac{x}{R} + \lambda_0$$

where R is the radius of the spherical reference surface; φ and λ are given in radians; l_0 is the central meridian of the projection; e = 2.7182818, the base of the natural logarithms, not the eccentricity.

Suppose a point, located at 60°N and 130°W, is projected on a map that uses the Mercator projection (where the reference surface is a sphere with a radius of 6371000 m. and the central meridian (*lo*) is 0°, equal to the Greenwich meridian). Using the forward mapping equation of the Mercator projection, the values found for the Cartesian coordinates are for x = -14,455,340m and for y = 8,390,339m.

The second example is the mapping equations used for the north polar *stereographic projection*:

The forward mapping equation is:

$$x = 2R\,tan\left(\frac{\pi}{4}-\frac{\phi}{2}\right)\sin(\lambda - \lambda_0)$$

$$y = -2R\,tan\left(\frac{\pi}{4}-\frac{\phi}{2}\right)\cos(\lambda - \lambda_0)$$

The inverse mapping equation is:

$$\phi = -2\arctan\left(\left(\sqrt{x^2+y^2}\,/\,2R\right)\right)+90^\circ$$

$$\lambda = \lambda_0 + \arctan\left(x/(-y)\right)$$

where R is the radius of the spherical reference surface; l_0 *is the longitude of the projection centre.*

Suppose the same point, located at 60°N and 130°W, is projected on a map that uses the stereographic projection (where the reference surface is a sphere with a radius of 6371000 m., the centre of the projection is located on the North pole and the longitude of the centre (l_0) is 0°). Using the forward mapping equation of the north polar stereographic projection, the values found for the Cartesian coordinates are for x = -2615435m and for y = 2194610m.

Types of Map Projection

Although the term 'projection' is used to describe the various transformations that enable us to view the spherical surface of the earth on a planar map, only a few actually are true 'projections' in the optical sense of the word. Indeed, there are two broad classes of "projections": true projections, and those based on mathematical transformations.

True Projections

These are formed by the projection of the earth's graticule of meridians and parallels on to some sort of screen or projection surface. Thus it is useful to imagine the earth as a transparent globe, perhaps made of clear glass, on which the meridians and parallels are painted as black lines. The position of the light source may be inside or outside the globe and of course the screen or projection surface can vary in shape as well.

Projection surfaces are of two general types: (a) a plane surface; and (b) developable surfaces. (a) The *plane surface* is the simplest projection surface and provides the most direct link between the global geography and the map. A plane can only touch the surface of a sphere at a single point of tangency and for this one unique point there is no distortion of the spherical grid on the map. For all other points, however, map distortion occurs and increases with distance from the point of tangency. Because directions from the point of tangency are shown correctly on a plane surface, these projections are also known as *azimuthal projections*. It is the type of projection surface usually used to map the polar regions of the planet.

b) *Developable surfaces* are projection surfaces which need to be 'unfolded' in order to produce a plane surface; there are two principal types: a cone and cylinder. The cone is developed by unrolling from a cut from the base to the apex of the cone; the resulting shape is a plane segment of a circle. The cylinder is developed by unrolling from a cut along the length of the cylinder; the resulting shape is a rectangle. The cone and cylinder form a continuum of geometry in the sense that a cylinder is a zero-angle cone.

Light-source locations for producing true projections commonly are fixed at

(a) the centre of the globe, called the *gnomonic position*;

(b) the antipode point of tangency of the projection surface, called the *stereographic position*; and

(c) at a point at infinite distance, called the *orthographic position.*

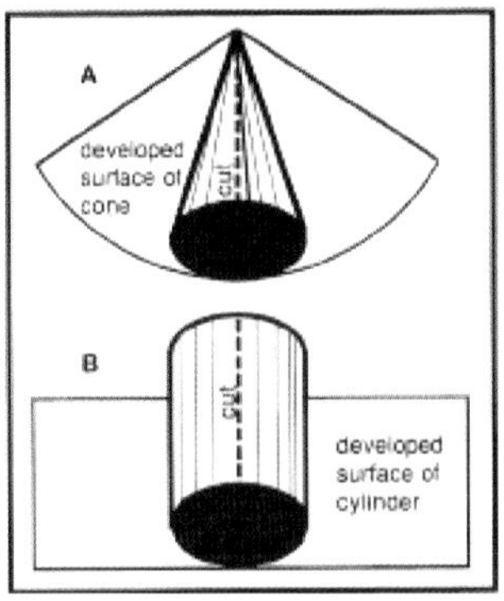

Developable surfaces
(A) a cone and (B) a cylinder

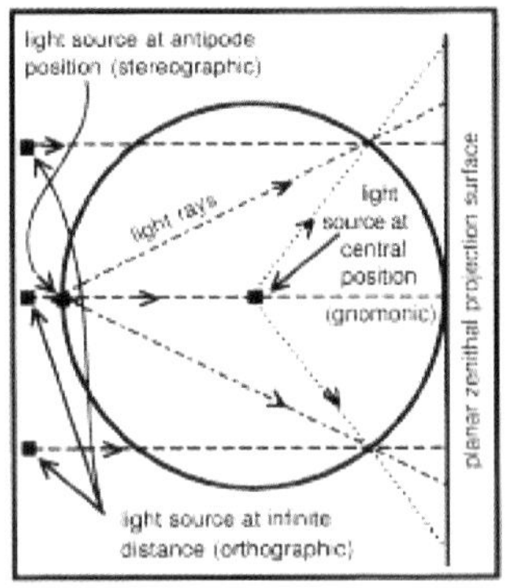

Three common light sources for true projections.

Orientation and tangency of projection surfaces also has an important effect on the character of the resulting map. It may be useful to think of the various possibilities here as being either (a) normal, (b) transverse; or (c) oblique.

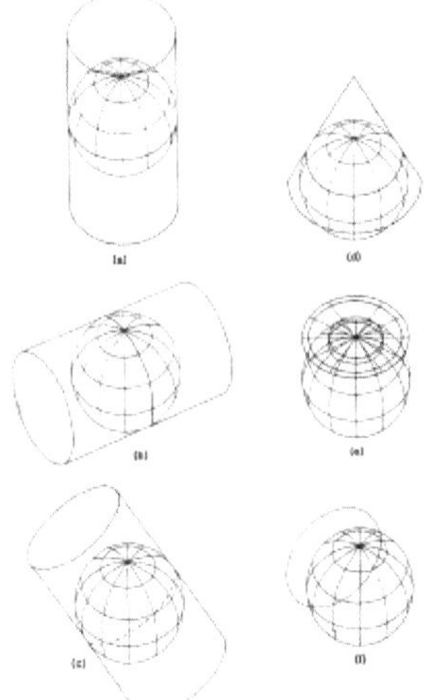

a) regular cylindrical; (b) transverse cylindrical; (c) oblique cylindrical; (d) regular conic; (e) polar azimuthal (plane); and (f) oblique azimuthal (plane).

(a) *Normal orientation and tangency* refers to those commonly used in mapping the globe on a particular kind of projection surface. For example, we already have noted that a plane projection surface tangent to the north or south geographic pole is a standard for mapping the Arctic and Antarctic, respectively. This choice of projection for the polar positions is not only recommended by its simplicity but it has the advantage of covering the land and ocean areas in question (within 30° of the polar tangency) with relatively little distortion. Cylindrical projections normally are employed so that they are tangent to Earth's equator. For this reason, only along the equator of a cylindrical projection are there no projection distortions. Thus, cylindrical projections in their normal orientation and tangency are ideally suited to the mapping of equatorial regions, especially within ±10°

strip about the equator, where the distortion of scale is less than +2% (distance between two points on the map parallel is 2% longer than the corresponding positions on the globe). The normal projection orientation of a cone is such that its axis and that of the globe's rotation is coincident. It is tangent to a parallel, higher in latitude as the cone angle increases. Thus the cone in its normal orientation and tangency is ideally suited to the mapping of middle latitudes. Cones, cylinders and planes are part of a geometric continuum, reflected in the fact that cones become planes and cylinders as the cone angle approaches its limits at 180° and 0°, respectively.

(b) *Transverse orientation* of the projection surfaces, as this class name suggests, involves a 90° rotation from the normal orientation. Thus, planes are tangent to the equator, cylinders are tangent to a meridian of longitude, and cones have axes passing through the equator.

(c) *Oblique orientation* refers to any intermediate orientation and tangency of the projection surface between the limiting classes of normal and transverse transformations.

Mathematical projections Not all 'projections' are true optical projections. Many are generated mathematically and the character of these transformations often are difficult to visualize. The spherical graticule of the globe is transformed in various ways to satisfy certain criteria and it will be easiest to examine this group of projections in terms of the few examples.

General Properties of Projections in Relation to the Globe

Because of its similarities to Earth, a sphere is a useful reference surface in any discussion of map projections. On a sphere, features of the earth's surface - their shape, area, and distance and directions between them - are shown correctly. The manner and degree to which these properties are distorted can be evaluated by reference to the spherical graticule forming the parallels of latitude and meridians of longitude. It might be useful to remind ourselves of the properties of this spherical grid system:

(a) Parallels of latitude measure an angle from Earth's centre in the polar plane, north and south with respect to the equator (0°); thus the geographic north and south poles are at respective latitudes 90°N and 90° S and the total latitudinal angular sweep from pole to pole is 180°.

(b) Parallels of latitude are indeed parallel, each with all others.

(c) Parallels of latitude are equally spaced; the polar circumference is approximately 40 008 km so that 1° of latitude corresponds with a spacing of $\frac{20\ 004\,\text{km}}{180^{\circ}}$ or 111.13 km on Earth's surface.

(d) Parallels of latitude are of unequal length. The longest parallel is the equator (40 076 km) and others decline in length north and south to a limiting point at the poles as follows:

0° = 40 076 km 20° = 37 674 km 40° = 30 743 km 60° = 20 084 km 80° = 6 982 km

10° = 39 471 km 30° = 34 736 km 50° = 25 812 km 70° = 13 748 km 90° = 0 km

(e) Meridians of longitude measure an angle from Earth's centre in the equatorial plane, 180° east and west with respect to the prime meridian (0°) through Greenwich, England; thus the total meridianal angular sweep is a full circle of 360°.

(f) All meridians pass through both poles and are equal in length (40 008 km).

(g) Meridians are not parallel and meridian spacing varies from 0 km where meridians meet at the poles to a maximum at the equator. 1° of equatorial longitude corresponds to a spacing of $\frac{40\ 076\,\text{km}}{360^{\circ}}$ or 111.32 km, while the corresponding spacing at intermediate latitudes of 20°, 40°, 60°, and 80°, for example, is respectively 104.65 km, 85.40 km, 55.79 km, and 19.39 km.

(h) Parallels of latitude and meridians of longitude everywhere cross at right angles.

An ideal map projection retains all these graticule characteristics through the translation to the map. But the ideal map projection can never be achieved and if a projection is designed to guarantee a particular characteristic as true, others on it necessarily will be distorted. As a result map projections will possess certain specific qualities of the globe but never all of them. In particular, map projections may be evaluated in terms of how well they preserve several global spatial dimensions: shape, area, distance, and direction.

Conformality is the term used to describe the property of correct shape retention on a map; such maps are said to be *orthomorphic*. The importance of conformality is that map features can be recognized by their distinctive shapes. If shapes are correct then directions must also be correct. It is necessary that, on all conformal maps, lines of latitude and longitude must cross at right angles and that the scale must be the same in all directions at any given point, just as is the case on the globe. Obviously the term conformal is somewhat misleading because there can be no such thing as a truly conformal map of the globe. But true shapes of small features may be retained on conformal map and the term is useful to denote this specific quality.

Equivalence refers to the retention of correct relative global areas on a map. Although it is possible to achieve equivalence in global map, such a projection always creates severe distortions in shape of the lands involved. To retain equivalence, any scale changes that occur in one direction must be offset by appropriate changes in the normal direction. Such projections obviously are important for maps depicting accurate relative areas.

Distance relationships on a map can be correctly depicted only if the length of a straight line between two points on the map projection represents the great circle distance between the same two points on the globe. It is only possible to depict correct distances on a map from one, or at most, two points. The azimuthal equidistant projection achieves this for all lines from the polar position (but for no others).

Direction is correctly retained on a projection when a straight line drawn between two points on the map shows the correct azimuth of the line; the azimuth is defined by the angle formed at the starting point of the straight line. In other words, a projection depicting true direction must show the great circle routes between points as straight lines. This is one of the properties of a gnomonic projection. Note that true direction or azimuth does not mean true bearing, an important distinction we will discuss later in the context of describing Mercator's projection.

Classification of Map Projections

Map projections are classified on the following criteria:

- Method of construction

- Development surface used
- Projection properties
- Position of light source

Method of Construction

The term map projection implies projecting the graticule of the earth onto a flat surface with the help of shadow cast. However, not all of the map projections are developed in this manner. Some projections are developed using mathematical calculations only. Given below are the projections that are based on the method of construction:

Perspective Projections : These projections are made with the help of shadow cast from an illuminated globe on to a developable surface

Non Perspective Projections :These projections do not use shadow cast from an illuminated globe on to a developable surface. A developable surface is only assumed to be covering the globe and the construction of projections is done using mathematical calculations.

Development Surface

Projection transforms the coordinates of earth on to a surface that can be flattened to a plane without distortion (shearing or stretching). Such a surface is called a developable surface. The three basic projections are based on the types of developable surface and are introduced below:

1. Cylindrical Projection

- It can be visualized as a cylinder wrapped around the globe.
- Once the graticule is projected onto the cylinder, the cylinder is opened to get a grid like pattern of latitudes and longitudes.
- The longitudes (meridians) and latitudes (parallels) appear as straight lines.
- Length of equator on the cylinder is equal to the length of the equator therefore is suitable for showing equatorial regions.

Aspects of cylindrical projection:

(a) (b) (c)

(a) Normal: when cylinder has line of tangency to the equator. It includes Equirectangular Projection, the Mercator projection, Lambert's Cylindrical Equal Area, Gall's Stereographic Cylindrical, and Miller cylindrical projection.

(b) Transverse: when cylinder has line of tangency to the meridian. It includes the Cassini Projection, Transverse Mercator, Transverse cylindrical Equal Area Projection, and Modified Transverse Mercator.

(c) Oblique: when cylinder has line of tangency to another point on the globe. It only consists of the Oblique Mercator projection.

2. Conic Projection

- It can be visualized as a cone placed on the globe, tangent to it at some parallel.
- After projecting the graticule on to the cone, the cone is cut along one of the meridian and unfolded. Parallels appear as arcs with a pole and meridians as straight lines that converge to the same point.
- It can represent only one hemisphere, at a time, northern or southern.
- Suitable for representing middle latitudes.

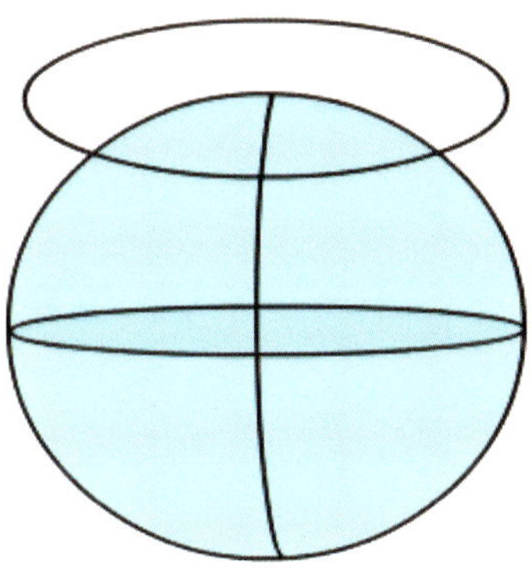

Aspects of conic projection:

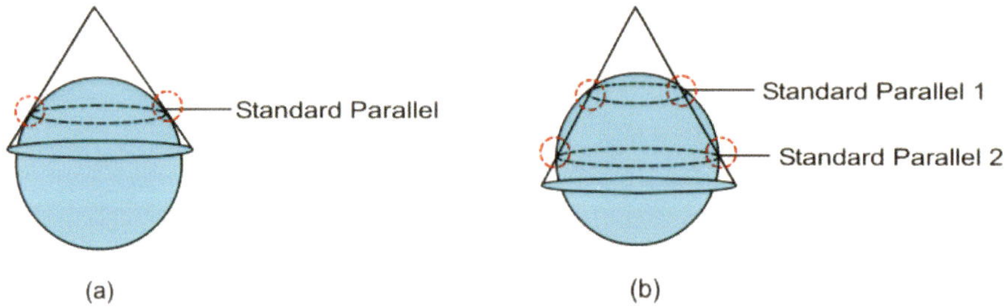

(a) Tangent: when the cone is tangent to only one of the parallel.

(b) Secant: when the cone is not big enough to cover the curvature of earth, it intersects the earth twice at two parallels.

3. Azimuthal/Zenithal Projection

- It can be visualized as a flat sheet of paper tangent to any point on the globe
- The sheet will have the tangent point as the centre of the circular map, where meridians passing through the centre are straight line and the parallels are seen as concentric circle.
- Suitable for showing polar areas

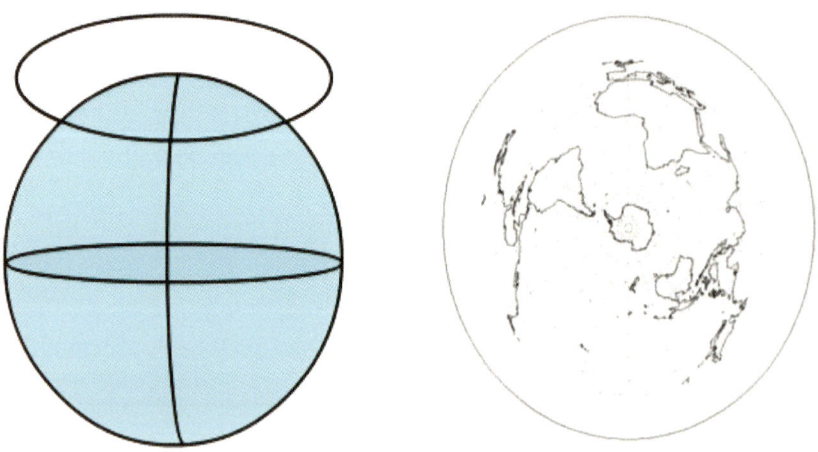

Aspects of zenithal projection:

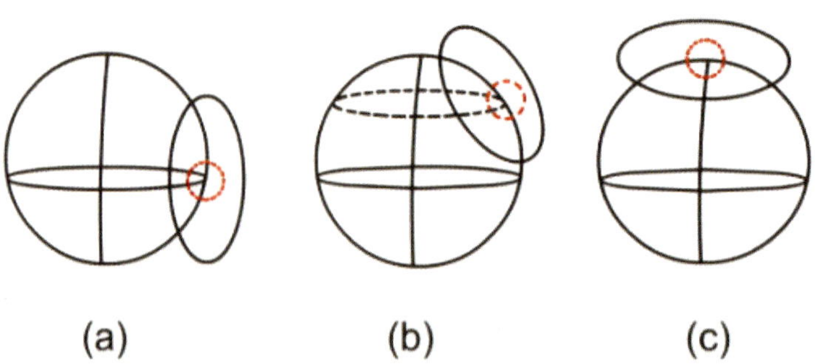

(a) Equatorial zenithal: When the plane is tangent to a point on the equator.

(b) Oblique zenithal: when the plane is tangent to a point between a pole and the equator.

(c) Polar zenithal: when the plane is tangent to one of the poles.

Projection Properties

According to properties map projections can be classified as:

Equal area projection: Also known as homolographic projections. The areas of different parts of earth are correctly represented by such projections.

True shape projection: Also known as orthomorphic projections. The shapes of different parts of earth are correctly represented on these projections.

True scale or equidistant projections: Projections that maintain correct scale are called true scale projections. However, no projection can maintain the correct scale throughout. Correct scale can only be maintained along some parallel or meridian.

Position of Light Source

Placing light source illuminating the globe at different positions results in the development of different projections. These projections are:

Gnomonic projection: when the source of light is placed at the centre of the globe.

Stereographic Projection: when the source of light is placed at the periphery of the globe, diametrically opposite to the point at which developable surface touches the globe.

Orthographic Projection: when the source of light is placed at infinity from the globe opposite to the point at which developable surface touches the globe.

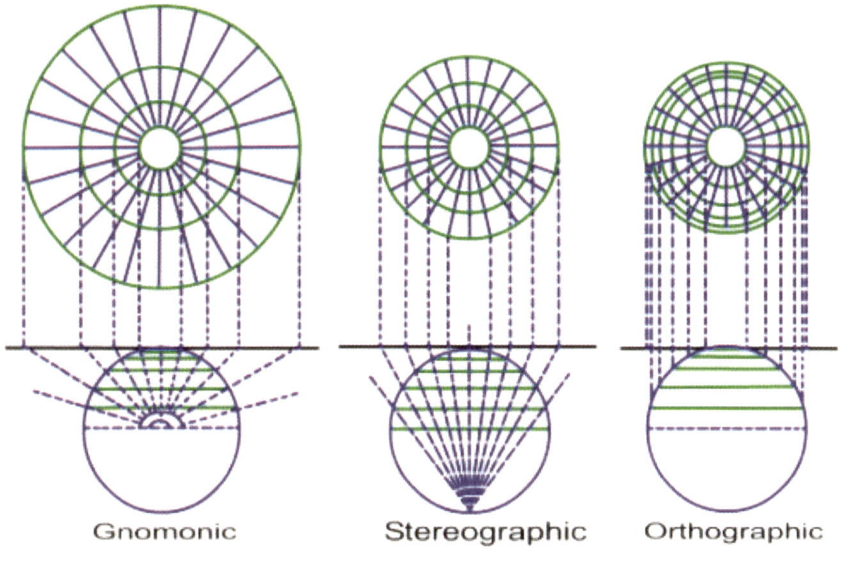

Projections and position of light source

Constructing Map Projections

Cylindrical Projection

Let us draw a network of Simple cylindrical Projection for the whole globe on the scale of 1: 400,000,000 spacing meridians and parallels at 30⁰ interval

$$\text{Radius of the earth} = 635{,}000{,}000 \text{ cm}$$

$$\text{Radius of globe on } 1:400{,}000{,}000 = 1/400000000 \times 635000000$$

$$= 1.587 \text{ cm}$$

$$\text{Length of the equator on the globe } = 2\pi, \text{ where } r \textit{ is radius of globe}$$

$$= 2 \times 22/7 \times 1.587$$

$$= 9.975 \text{ cm}$$

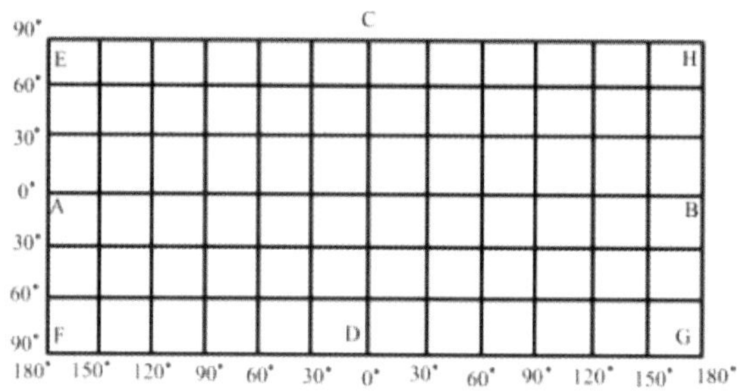

Simple cylindrical projection graticule

Steps of Constrution:

- Draw a line AB, 9.975 cm long to represent the equator. The equator is a circle on the globe and is subtended by 360⁰.
- Since the meridians are to be drawn at an interval of 30⁰ divide AB into 360/30 or 12 equal parts.
- The length of a meridian is equal to half the length of the equator i.e. 9.975/2 or 4.987 cm.
- To draw meridians, erect perpendiculars on the points of divisions of AB. Take these perpendiculars equal to the length specified for a meridian and keep half of their length on either side of the equator.
- A meridian on a globe is subtended by 180⁰. Since the parallels are to be drawn at an interval of 30⁰, divide the central meridian into 180/30 i.e. 6 parts.
- Through these points of divisions draw lines parallel to the equator. These lines will be parallels of latitude. Mark the equator and the central meridian with 0⁰ and the parallels and other meridians. EFGH is the required graticule.

Conical Projection

Let us draw a graticule on simple conical projection with one standard parallel on the scale of 1: 180,000,000 for the area extending from the equator to 90⁰ N latitude and from 60⁰ W longitude to 100° E longitude with parallels spaced at 15° interval, meridians at 20°, and standard parallel 45° N.

Calculations:

$$\text{Radius of the earth} = 635{,}000{,}000 \text{ cm}$$

$$\text{Therefore, radius of the globe on the scale of } 1:180{,}000{,}000$$

$$= 1/180{,}000{,}000 \times 635{,}000{,}00$$

$$= 3.527 \text{ cm}$$

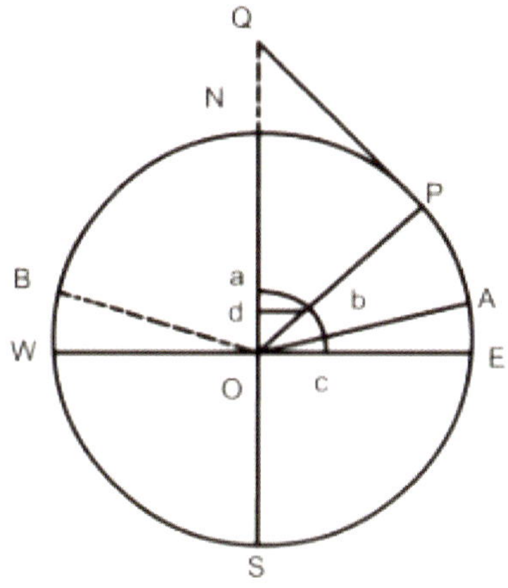

Steps of construction:

1. Draw a circle with a radius of 3.527 cm that represents the globe. Let NS be the polar diameter and WE be the equatorial diameter which intersect each other at right angles at O.
2. To draw the standard parallel 45° N, draw OP making an angle of 45° with OE.
3. Draw QP tangent to OP and extend ON to meet PQ at point Q.
4. Draw OA making an angle equal to the parallel interval i.e. 15° with OE.

Draw line LM, it represents the central meridian

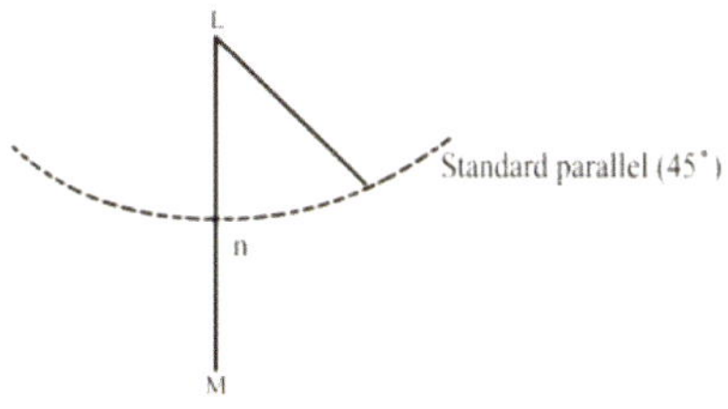

With L as the centre and QP as the radius, draw an arc intersecting LM at n. This arc describes the standard parallel 45° N.

The distance between the successive parallels is 15°. The length of the arc subtended by 15° is calculated as under:

$$2\pi \times 1/360 \times 15$$
$$= 2 \times 22/7 \times 3.527 \times 15/360$$
$$= 0.923\,\text{cm}$$

From point n, mark off distances nr, rs, st, nu, uv and vM , each distance being equal to 0.923 cm. With L as centre, draw arcs passing through the points t, s, r, u, v and M. These arcs represent the parallels.

1. Draw OB making an angle of 20° with OW Length of the arc subtended by 20° is calculated as under:

$$2\Pi r \times 1/360 \times 20$$
$$= 2 \times 22/7 \times 3.527 \times 20/360$$
$$= 1.231\ \text{cm}$$

1. With O as centre and radius equal to the arc WB (1.231 cm) draw arc abc.
2. From point b, drop perpendicular bd on line ON. Now db is the distance between the meridians.

Keeping in view the number of meridians to be drawn, mark off distances along the standard parallel toward the east and west of the point n, each distance being equal to db.

Join point L with the points of divisions marked on the standard parallel and produce them to meet the equator.

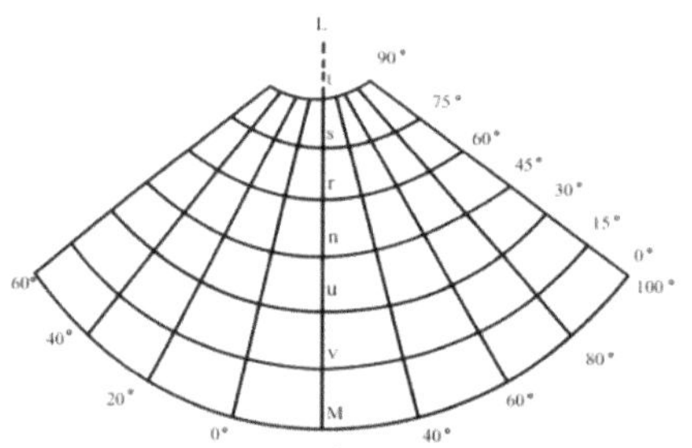

Simple conic projection

Azimuthal Projection

Let us draw Polar zenithal equal area projection for the northern hemisphere on the scale of 1: 200,000,000 spacing parallels at 15° interval and meridians at 30° interval.

Calculations:

$$\text{Radius of the earth} = 635{,}000{,}000\ \text{cm}$$
$$\text{Therefore, radius of the globe on the scale of } 1:200{,}000{,}000$$
$$= 1/200{,}000{,}000 \times 635{,}000{,}000$$
$$= 3.175\ \text{cm}$$

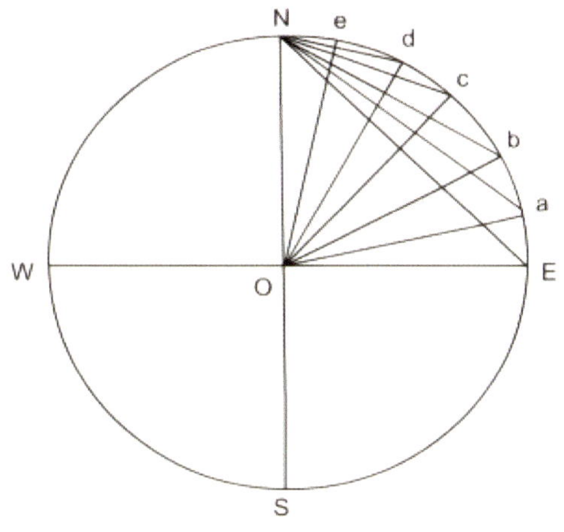

Steps of construction:

- Draw a circle with radius equal to 3.175 cm representing a globe. Let NS and WE be the polar and equatorial diameter respectively which intersect each other at right angles at O, the centre of the circle.
- Draw radii Oa, Ob, Oc, Od, and Oe making angles of 15°, 30°, 45°, 60° and 75° respectively with OE. Join Ne, Nd, Nc, Nb, Na and NE by straight lines.
- With radius equal to Ne, and N' as centre draw a circle. This circle represents 75° parallel. Similarly with centre N' and radii equal to Nd, Nc, Nb, Na and NE draw circles to represent the parallels of 60°, 45°, 30°, 15° and 0° respectively.
- Draw straight lines AB and CD intersecting each other at the centre i.e. point N.
- Radius N'B represents 0° meridian, N'A 180° meridian, N'D 90° E meridian and N'C 90° W meridian.
- Using protractor, draw other radii at 30° interval to represent other meridians.

Commonly used Map Projections

Rather than attempt a comprehensive cataloguing of the many different types of map projections in use, here we will simply examine the character of several examples which are either commonly usedmap projections, or have particularly instructive properties.

The projections considered here illustrate well the general problem of transforming a spherical surface to a plane. Insights gained from evaluating these examples will help in assessing other map projections we might encounter in the future. Our examples constitute four groups, (a) cylindrical-type projections, (b) a compromise projection (Mollweide's), (c) zenithal projections, and (d) conical-type projections:

(a) Cylindrical-type projections

1. Perspective cylindrical projection
2. Plate carrée (or simple cylindrical or cylindrical equidistant projection

3. Cylindrical equal area projection
4. Mercator's projection (cylindrical orthomorphic)
5. Sinusoidal projection
6. Interrupted sinusoidal

(b) A compromise projection

7. Mollweide's projection

(c) Zenithal projections

8. Gnomonic projection
9. Stereographic projection
10. Orthographic projection
11. Zenithal equidistant projection
12. Zenithal equal-area projection

(d) Conical-type projections

13. Perspective conic projection
14. Simple conic projection: one standard parallel
15. Bonne's projection
16. Conic projection with two standard parallels
17. Polyconic projection

Cylindrical-type projections

The perspective cylindrical projection can be thought of as the basic model from which all the projections in the first group are derived. It is a true projection formed by projecting a globe from its centre onto a circumscribing cylinder. Beyond providing a basis for modification, however, it is of little practical use because of the obviously extreme scale-distortion at high latitudes; clearly the parallel length and meridian spacing should decline towards the poles. The only redeeming feature of this projection is that, within 100 of the equator, scale distortion is less than a few per cent. Because it can be readily improved, other derivative projections are distinctly superior.

The plate carrée (or simple cylindrical or cylindrical equidistant) projection is not a true projection but we can think of it as a cylindrical-type projection. It is a rectilinear graticule in which the meridians are true length and parallels all equal the length of the equator and are correctly spaced as on the globe. As in the case of all cylindrical-type projections, however, because the meridian spacing does not decline toward the poles, the polar regions consequently are highly distorted. Nevertheless, although neither shape nor area can be correctly depicted on this projection, within a 5° strip on either side of the equator, exaggeration along the parallels is less than +0.38%.

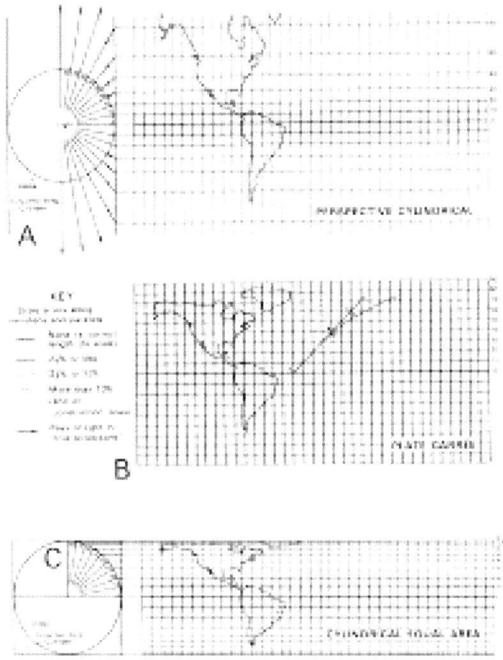

Cylindrical-type projections (developed from straight lines). Projections C, D, E, and F can all be regarded as modifications of the basic network of either the plate carrée (B) or the perspective cylindrical projection (A); scale errors have been generalized to some extent.

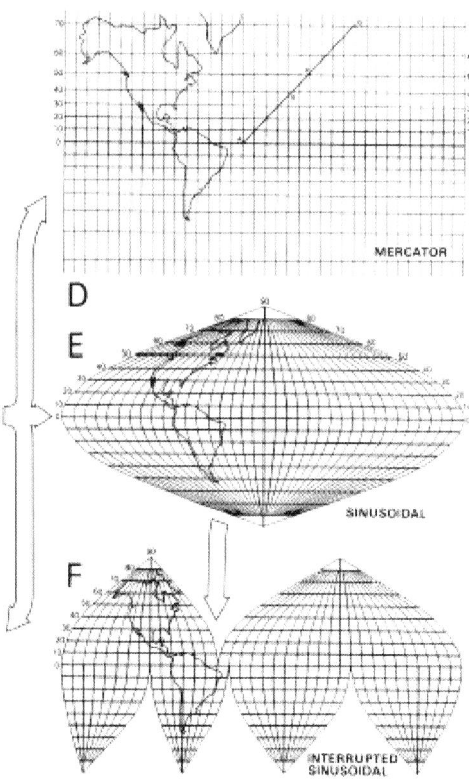

(continued): Note that a rhumb line AEFG on Mercator's projection (D) is represented by a curved course on the plate carrée (B) and that a straight line ABCD (i.e., an apparently fixed heading) on the plate carrée deviates from a true rhumb line (after Dickinson, G.C.,1969, Maps and Air Photographs. 2nd ed. New York: John Wiley and Sons; with permission).

The plate carrée is of great historical significance because it was invented by Eratosthenes some two thousand years ago and has been used as a map base at various times since then until quite recently. For example, it was used for the early editions of the Ordnance Survey maps of England and Wales.

The cylindrical equal-area projection is a true projection in which a light from a source in the orthographic position (at infinite distance) is projected onto a cylindrical surface in the normal orientation and tangency. Alternatively, we can think of the cylindrical equal-area projection as a perspective cylindrical (or plate carrée) projection in which the spacing of parallels has been adjusted so that relative areas are shown correctly as they are on the globe. As the parallels are stretched beyond their true length toward the poles on this rectilinear graticule, the parallel spacing must be compressed in compensation in order to preserve equal areas. Thus, the cost of obtaining equality of area on this projection is gross scale distortion in high latitudes. In fact it turns out that there are better solutions to this equal-area problem than that offered by the cylindrical equal-area projection and it is little used in modern cartography.

Mercator's projection (cylindrical orthomorphic) probably has been and remains the most widely used (and abused) map projection during the last 400 years. The Flemish geographer and mathematician Gerardus Mercator (1512-1594) designed his projection in 1569 as a navigational aid for sailors. He wanted a map on which a compass heading or line of constant bearing (rhumb line) would appear as a straight line. This property would allow navigators to transfer compass headings directly to the map and to directly measure course headings between points shown on the map. The Mercator projection, like the other projections in this group, is a derivative of the perspective cylindrical or plate carrée projections. Like the plate carrée it is based on a rectilinear graticule in which parallels are drawn the same length as the equator throughout and meridian spacing corresponds to that on the equator. In order to satisfy this particular navigational requirement, however, the parallel spacing (meridian length) increases away from the equator at the same rate as the parallels lengthen compared with their true length on the globe. Mercator's pre-calculus solution to this problem of continuous change was only approximate but it was sufficiently accurate to be of great practical use to mariners.

The distinction between *great circles* and *rhumb lines* is important to appreciating the character of Mercator's projection. If you trace out the shortest line between two points on the surface of the globe you have traced a so-called *great circle route*. It is a segment of a circle whose centre is the centre of the globe. For example, consider the great circle route between Vancouver, British Columbia and London England. Because this is the shortest distance between these two points it is the non-stop route followed by commercial airlines. Such a flight leaves Vancouver at a true northeast bearing (about 45°) but approaches London on a southeast heading (about 135°). Midway in the flight, over Baffin Island, the aircraft would have been flying parallel to the Arctic Circle (66.6°N) on a true east-west bearing of 90°. In other words, because the great circle route diagonally crosses the spherical graticule, the bearing the aircraft follows (as measured by an onboard magnetic compass) in order to keep to the great circle route, is constantly changing.

The initial bearing from one point to another (450 at Vancouver in our example) is known as an *azimuth*. It will be clear that flying a course set at the azimuth of 450 from Vancouver will not get our aircraft to London! Such a line of constant bearing is known as a *rhumb line*. A 450 rhumb line from Vancouver would take us on a curved course, spiraling in to the North Pole.

The problem facing Mercator was to produce a map which would show a straight line between any two points as a rhumb line. The great utility of such a map is that a sailor could draw a straight line between his planned point of departure and his destination and measure directly the compass heading he would need to follow in order to successfully complete his journey. On Mercator's projection an azimuth lies on a rhumb line. Since rhumb lines are not great circle routes (except for the special cases of a route exactly on the equator or exactly on a meridian), it follows that a rhumb line course plotted on Mercator's projection is not the shortest route. Conversely, great circle routes plotted on Mercator's projection appear as curved lines. For example, a straight line between Vancouver and London on Mercator's projection is an approximately east-west (90°) rhumb line which takes us on a southerly course over Winnipeg and Newfoundland, and is about 20 per cent longer than the great circle route over Greenland.

But Mercator made no claim for economy; rather he claimed, 'a straight line between two points on Mercator's projection is a rhumb line and if you keep your compass fixed to the rhumb line azimuth you will reach your destination'. A navigator wanting both economy of travel and ease of navigation must use *two* map projections. First, the route must be plotted on a gnomonic projection on which all straight lines are great circle routes (but not rhumb lines). This route must then be transferred to Mercator's projection where it will plot as a curved line. This arc can be broken up into a series of connected straight chords or legs and thus each leg is a rhumb line yielding a compass heading. By sailing along this series of rhumb lines a navigator can approximate the shortest route.

Scale distortion on Mercator's projection is very severe at high latitudes but quite tolerable (less than +1.5%) within 10° of the equator. Above 60° latitude scale errors exceed +100% and become infinite at the poles; we are all familiar with the typical school atlas showing Greenland as large as South America when in fact it is only about one tenth of that continent's area! The alternative name for Mercator's projection, the *cylindrical orthomorphic* projection, recognizes that it fulfills the property of orthomorphism so that shapes over very small areas therefore are shown correctly.

Given its intended specialized navigational use, it is rather surprising how widely Mercator's projection has been used as a basis for world maps. No doubt part of the explanation lies in the fact that for many years in many places navigational charts were the most readily available maps, and indeed the only ones available, and the inertia of convention has helped to keep it in use. In its transverse form it is of course the projection (UTM) used here in Canada for the NTS. The sinusoidal projection, also known as the Sanson-Flamsteed projection, is derived directly from the plate carrée by adjusting the equatorial length of parallels, and therefore the spacing of the meridians, so that they are equal to their true values for the globe. The result is a rather aesthetically displeasing ellipsoidal outline pinched to a peak at both poles. Nevertheless, it has the virtue of a central region of minimal distortion rather than an equatorial strip as in the case of the plate carrée, for example. As we move away from the equator and central meridian of this projection, however, the graticule intersections increasingly skew from 90° so that in the peripheral areas shapes are highly distorted. Still, compared with the perspective cylindrical projection, plate carrée, or Mercator's projection, high latitude shapes and areas are far less distorted. In fact, the sinusoidal projection retains the property of equivalence over the entire projection. This property follows from the theorem showing that parallelograms with similar bases drawn between the same parallel lines are equal in area regardless of shape. As an equal-area projection the sinusoidal projection is better than a cylindrical equal-area projection which has extreme shape distortion at high latitudes.

A solution to the shape-distortion problem on the sinusoidal and certain other projections is to interrupt the graticule along selected meridians. An interrupted version of the sinusoidal projection is shown in Figure. Here several central meridians of low distortion are employed along with graticule breaks so that the overall shape distortion is reduced. But the result clearly is rather odd! Now we have a map which appears to have been torn into strips and furthermore it cannot be reformed to an uninterrupted version by cutting and pasting; adjoining sheets do not match well without boundary distortion. Nevertheless, if we are just interested in seeing well-shaped equal-area continental areas on the one projection, interruption can be confined to the oceans so that information loss and distortion is minimal. Certainly you will find examples of interrupted projections in most world atlases.

(b) A compromise projection

Mollweide's projection, also known as the homolographic projection, is a very commonly used example of a mathematical projection developed to correct some of the worst features of shape distortion in equal-area projections such as the cylindrical equal-area and sinusoidal projections. It is simply a mathematical response to the need for an equal-area projection formed within an elliptical frame in which the parallels are straight lines; the central meridian is one half the length of the equator.

Although Mollweide's projection also suffers from the skewing of meridians as they are gathered together at the poles, it distributes shape distortions more evenly over the map and provides a more aesthetically pleasing map than does the plate carrée or sinusoidal projections. The Mollweide projection is widely used in atlases for mapping areal data and like the sinusoidal projection, Mollweide's often is presented in interrupted form to reduce overall shape distortion on single-sheet world maps.

We might note here that cartographer J.P. Goode has developed a hybrid projection (the *homolosine* projection) formed by a sinusoidal projection from the equator to 40° and capped by Mollweide's projection in the higher latitude areas, thus retaining the best of both components and minimizing distortion overall.

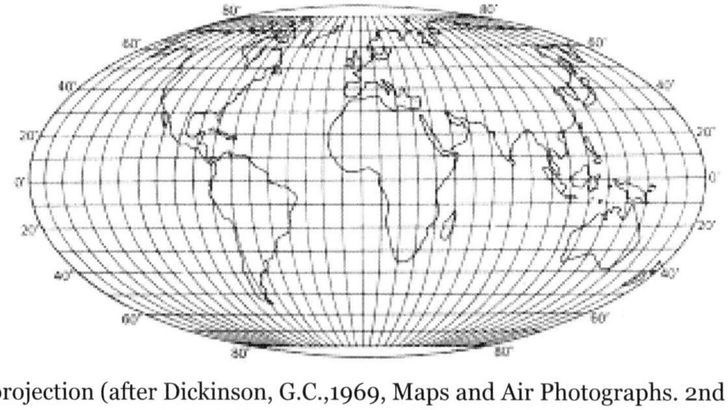

Mollweide's projection (after Dickinson, G.C.,1969, Maps and Air Photographs. 2nd ed. New York: John Wiley and Sons; with permission).

(c) Zenithal projections

The gnomonic projection is one of several types of true projection in which the spherical graticule is projected onto a plane surface touching the globe at a single point of tangency. In particular, the gnomonic projection is a true projections from a light source at the earth's centre. The projected graticule is limited to a hemisphere or less and the outer areas are severely distorted. It does have the rare property, however, that, because light is originating from the earth's centre of radius, any straight line drawn on a gnomonic projection is part of a great circle and thus the map is very useful for navigation when used in conjunction with the Mercator projection. Bearings are azimuths when measured from the centre (polar position) of these maps.

Other true projections closely related to the gnomonic projection are the *orthographic* and *stereographic* projections. The orthographic projection is simply the globe seen from afar so it is perhaps the most familiar and therefore easiest projection to envisage. The stereographic projection is better balanced than the orthographic and gnomonic projections in the sense that scale distortion is more evenly distributed over the graticule. For example, scale errors are less than +7.2% up to 40° from the point of tangency compared with +30.5% for 40° from the equator on Mercator's projection or the plate carrée. Indeed, errors do not reach +33.3% on the stereographic projection until 70° from the point of tangency. Although it is not readily be apparent, the stereographic projection also is an orthomorphic projection but unlike Mercator's it is not in general use for topographic maps (the Dutch official map series provides an exception).

The zenithal equidistant and equal - area projections are mathematical derivatives of the

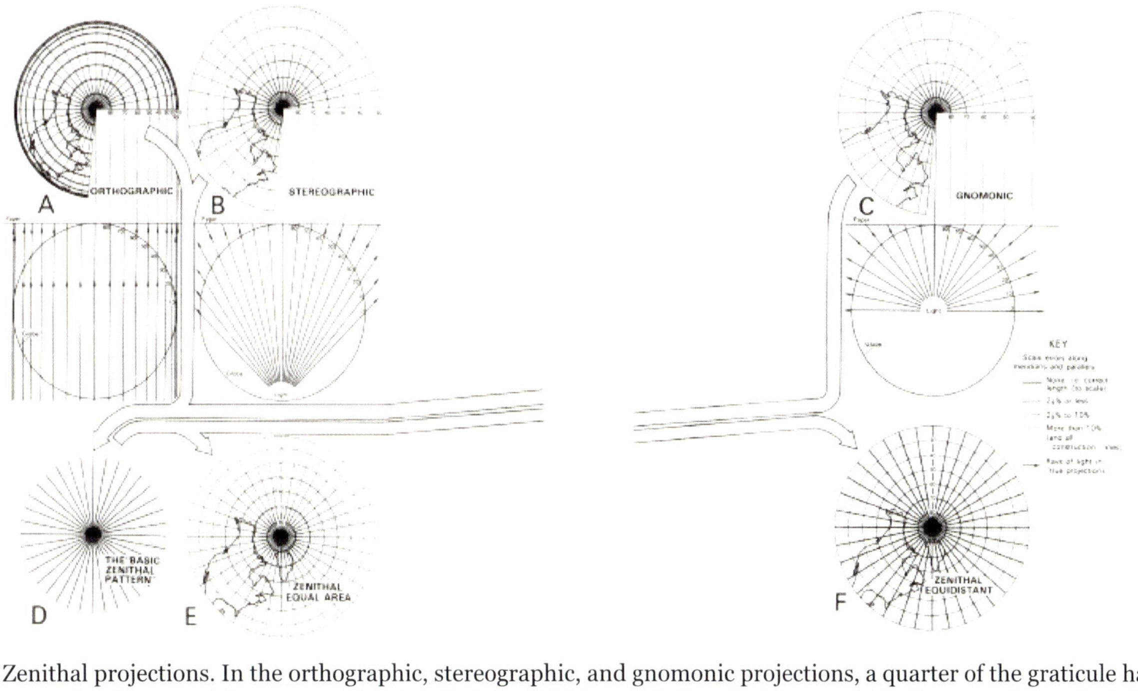

Zenithal projections. In the orthographic, stereographic, and gnomonic projections, a quarter of the graticule has been omitted to reveal the construction details (after Dickinson, G.C.,1969, Maps and Air Photographs. 2nd ed. New York: John Wiley and Sons; with permission).

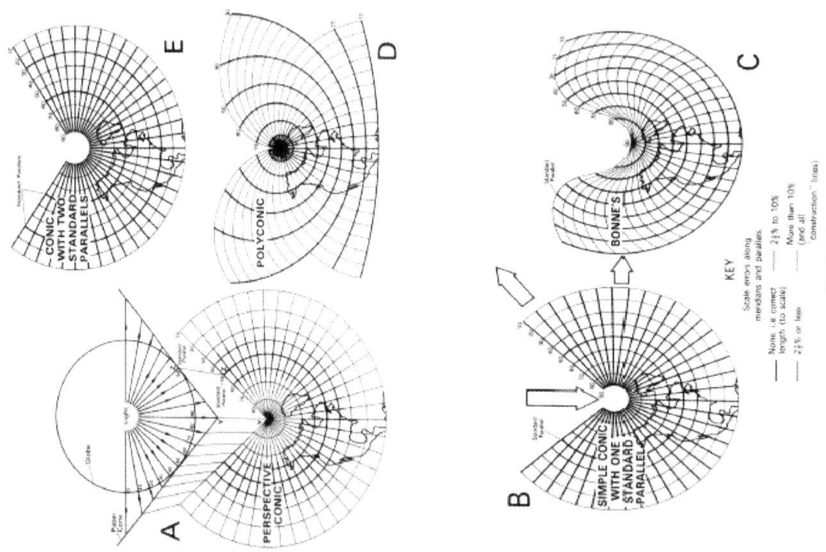

orthographic-stereographic-gnomonic family of true projections. In its conventional polar tangency the basic zenithal projection consists of straight meridians radiating from the pole which intersect concentric parallels at right angles. In the zenithal equidistant projection the parallel spacing is modified to show true distance along the meridians (but not in any other direction). The zenithal projections are very good for depicting polar regions (even up to 45° scale distortions are less than about +10%) and in their transverse or oblique forms they are usefully applied in other areas as well.

The zenithal equal-area projection is constructed by adjusting the spacing of parallels of the basic zenithal projection so that they enclose on the map the same area as they do on the globe, thus retaining the property of equivalence Figure). It turns out that shape distortion is also minimal (less than about +10%) within 50° of the pole as well.

As in all zenithal projections, directions from the centre of this projection are shown correctly as azimuths of great circles. Zenithal projections are not widely used in topographic mapping, however, because the conicals do an even better job of portraying small areas.

(d) Conical-type projections

The perspective conic projection is the basic form of the final group of projections to be considered here: those constructed on a conical surface. Although only one cylinder will fit the globe, and a flat projection surface has fixed geometry, any number of different cones can rest on the globe, from wide-angle types touching near the poles to low-angle cones touching near the equator. Indeed, as we noted earlier the plane and cylinder can be thought of as limiting cones. The cone angle is selected so that the circular contact between the cone and the globe best suits the area being mapped. For example, the projections shown in Figure make contact at 50° N because this tangency suits maps of North America. In the normal orientation this line of tangency defines the *standard parallel* of the projection.

As with cylindrical projections, the true or perspective conic projection is of little practical use but it does provide a basic model for modification. The perspective conic projection has a standard parallel shown at its correct length as an arc whose radius is the slant height of the cone at that point. Other parallels are arcs of concentric circles; meridians are straight lines radiating from the centre of these circles through points correctly spaced along the standard parallel. True distances along the parallels are exaggerated away from the standard parallel (i.e., in both radial directions). Parallels and meridians intersect at right angles, however, so we might expect minimal shape distortion.

The simple conic projection with one standard parallel is a mathematical projection derived from the perspective conic. In this case, parallels are spaced at their true distance along the meridians and the pole becomes a small arc rather than a point. As with the perspective conic projection scale is exaggerated along all parallels other than the standard. Scale errors and thus distortions in areas and shape are very small in a narrow band about the standard parallel.

Bonne's projection is a derivative of the simple conic with one standard parallel. It has been modified to remove the exaggeration of scale along the parallels by redrawing them at their correct length. In consequence, meridians become curved lines and no longer intersect parallels at right angles. Furthermore, meridians have been slightly stretched to longer than actual length. Nevertheless, in spite of these shape distortions, Bonne's projection is an equal area projection and shapes and errors are reasonably controlled about the central meridian. The projection is widely used in topographic surveys and in atlas presentations.

The conic projection with two standard parallels, although essentially a mathematical construct, can be envisaged as a derivative of the perspective conic. We can think of it as a spherical graticule projected onto a cone which, instead of touching the globe at a point of tangency, intersects the global surface at two parallels. These two standard parallels are then adjusted so that they are correctly spaced as on the globe. Meridians are correctly spaced along the arcs of the two standard parallels and radiate from the centre of the arcs. Other parallels are correctly spaced arcs concentric to the standard parallels. The resulting graticule has properties similar to those of the simple conic projection with one standard parallel except that errors are more evenly distributed across the map. Scale is correct along all meridians and along both standard parallels (slightly too small between them and too large along parallels outside the standards). Errors are quite small over large areas; for example, the whole of U.S.A. can be mapped with less than 2% scale error.

The polyconic projection provides a means of extending the coverage of the conic projection beyond a single hemisphere. The simple conic with one standard parallel set to accurately portray the mid-latitudes of one hemisphere necessarily will produce a severely distorted and unacceptable view of the other hemisphere. Employment of a second standard parallel can take our projection domain somewhat beyond the equator with less overall distortion but the cost is increased specific distortion in the polar region as well as in the mid latitudes, compared with that associated with a simple conic with a standard parallel tangent in these regions. The polyconic projection is the logical extension of the error distributing feature of the conic projection with two standard parallels. If two standard parallels generally are better than one, then why not employ three or ten or one hundred? Well, obviously a single cone cannot intersect the surface of a globe more than twice (to yield two standard parallels) but we can solve this problem by using more than one cone. Indeed, in the polyconic projection every parallel is a standard parallel and although it is a mathematical projection it can be conceived as being derived from an infinite number of cones. Because the cone angle changes for each of these standard parallels they are no longer concentric but they are spaced correctly along a central meridian and the meridians also are correctly spaced along the parallels. North-south scale errors increase rapidly away from the central meridian so that the projection is of little use as a global map in the form shown in Figure. It is widely used in its interrupted form, however, with pole to pole mapping restricted to narrow strips centred on the central meridian. For example, it is projection used in the International 1:1 000 000 map of

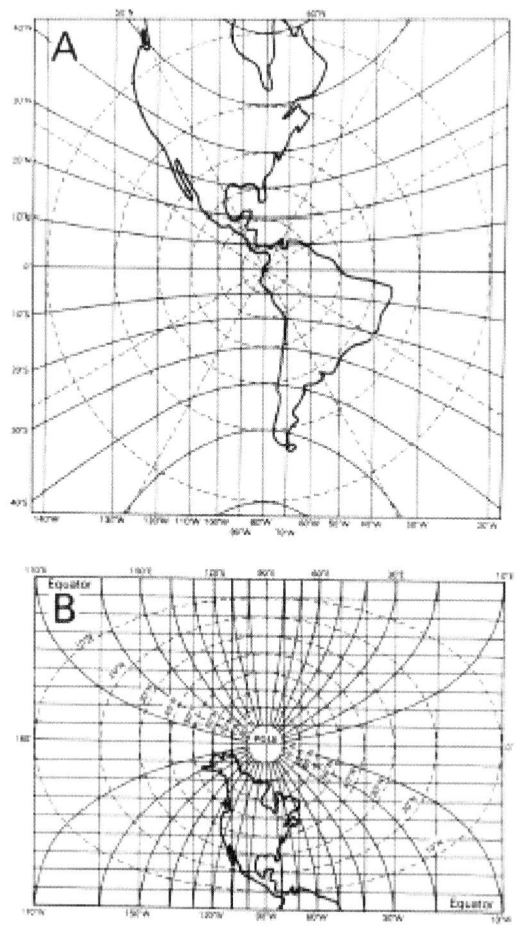

Transverse versions of projections. A: Transverse gnomonic projection centred on the equator at 80o W, and B: Transverse Mercator projection, central meridian 90o W (only one hemisphere). The pecked lines in (A) and faint lines in (B) indicate the patternsof meridians and parallels in the normal versions (after Dickinson, G.C.,1969, *Maps and Air Photographs.* 2nd ed. New York: John Wiley and Sons; with permission).
the world and in the maps of the United States Geological Survey.

Each of the above projections has been described in terms of its normal orientation and tangency but of course every one can be rotated to any number of oblique positions or to a full transverse orientation. These rotated projections often result in an unfamiliar pattern of parallels and meridians but the properties of scale distortion of the projections are exactly the same as those in the normal orientation. Figure shows the graticule for the transverse Mercator projection (in this case, centred on meridian 90oW), the basis of the National Topographic Series (NTS) here in Canada.

Selection of Map Projection

Choosing a correct map projection for an area depends on the following:

Map Purpose

Considering the purpose of the map is important while choosing the map projection. If a map has a specific purpose, one may need to preserve a certain property such as shape, area or direction

On the basis of the property preserved, maps can be categorized as following

a. Maps that preserve shapes.

Used for showing local directions and representing the shapes of the features. Such maps include:

- Topographic and cadastral maps.
- Navigation charts (for plotting course bearings and wind direction).
- Civil engineering maps and military maps.
- Weather maps (for showing the local direction in which weather systems are moving).

b. Maps that preserve area

The size of any area on the map is in true proportion to its size on the earth. Such projections can be used to show

- Density of an attribute e.g. population density with dots
- Spatial extent of a categorical attribute e.g. land use maps
- Quantitative attributes by area e.g. Gross Domestic Product by country
- World political maps to correct popular misconceptions about the relative sizes of countries.

c. Maps that preserve scale

Preserves true scale from a single point to all other points on the map. The maps that use this property include:

- Maps of airline distances from a single city to several other cities
- Seismic maps showing distances from the epicenter of an earthquake
- Maps used to calculate ranges; for example, the cruising ranges of airplanes or the habitats of animal species

d. Maps that preserve direction

On any Azimuthal projection, all azimuths, or directions, are true from a single specified point to all other points on the map. On a conformal projection, directions are locally true, but are distorted with distance.

General Purpose Maps

There are many projections which show the world with a balanced distortion of shape and area. Few of these are Winkel Tripel, Robinson and Miller Cylindrical.For larger-scale maps, from continents to large countries, equidistant projections are good at balancing shape and area distortion. Depending on the area of interest, one might use Azimuthal Equidistant,Equidistant Conic and Plate Carrée.

References

- Kurt A. Raaflaub; Richard J. A. Talbert (2009). Geography and Ethnography: Perceptions of the World in Pre-Modern Societies. John Wiley & Sons. p. 147. ISBN 1-4051-9146-5
- Catherine Delano Smith (1996). "Imago Mundi's Logo the Babylonian Map of the World". Imago Mundi. 48: 209–211. doi:10.1080/03085699608592846. JSTOR 1151277
- Sun, Shipeng (2013), "A Fast, Free-Form Rubber-Sheet Algorithm for Contiguous Area Cartograms", International Journal of Geographic Information Science, 27 (3): 567–93, doi:10.1080/13658816.2012.709247
- Robinson, A.H. (1982). Early Thematic Mapping: In the History of Cartography. Chicago: The University of Chicago Press. ISBN 0-226-72285-6
- Jenny, Bernhard; Hurni, Lorenz (2006). "Swiss-Style Colour Relief Shading Modulated by Elevation and by Exposure to Illumination". The Cartographic Journal. 43 (3): 198–207. doi:10.1179/000870406X158164
- "Maclure's geological map of the United States". US Library of Congress' Map Collection. Library of Congress. Retrieved 30 October 2015
- Jenny, Bernhard; Hurni, Lorenz (2006). "Swiss-Style Colour Relief Shading Modulated by Elevation and by Exposure to Illumination". The Cartographic Journal. 43 (3): 198–207. doi:10.1179/000870406X158164
- Needham, Joseph (1971). Part 3: Civil Engineering and Nautics. Science and Civilization in China. 4. Cambridge University Press. p. 569. ISBN 978-0-521-07060-7
- Sun, Shipeng (2013), "An Optimized Rubber-Sheet Algorithm for Continuous Area Cartograms", The Professional Geographer, 16 (1): 16–30, doi:10.1080/00330124.2011.639613
- "Nepal-Topo Maps". PAHAR Mountains of Central Asia Digital Dataset. MyPublisher. Archived from the original on 5 January 2014. Retrieved 31 January 2014

5

Understanding Geographical Databases

A geographical database is a database that stores geographical data pertaining to administrative divisions, cities and countries. This chapter has been carefully written to provide an easy understanding of the varied facets of geographical databases, by incorporating several significant topics such as aerodrome mapping database, barrow area information database, European soil database, etc.

A Geographic database can be defined as a catalog that stores data that are spatially referenced. These databases are collections of similar data that are related either through location, data structure or type, or common underlying purpose.

Geographic databases can be divided into two main data concepts, graphic and nongraphic. Nongraphic data is data that does not "describe" or "define" graphic images or map features. There are four types of nongraphic data: attributes, geographically referenced data, spatial relationships, and geographic indexes. Generally, nongraphic data attributes describe particular map features or are linked to graphic elements through identifiers or geocodes. An example of an identifier or geocodes would be used to describe items or actions at a geographic location such as building permits, accident reports, or tax records. Nongraphic data within a geographic database can take the form of a geographic index or be used to describe a spatial relationship.

Examples of geographic indexes include street addresses, mailing addresses, parcel numbers, or account numbers. Spatial relationship examples include various topological relationships such as connectivity, adjacency, and proximity. All of these elements of nongraphic geographic data are stored in geographic databases as alphanumeric characters.

Graphic data in geographic databases consists of points, lines, polygons, and other map or cartographic features such as projections, coordinate systems, and cartographic symbols. Graphic data is commonly stored in the database in the form of coordinates, symbols, rules, or pixels. There are two methods in which graphic data can be stored within a geographic database: vector or raster. Vector data are denoted by coordinates (example, x and y or latitude and longitude) of nodes and lines or rules for connecting the various lines into areas. Vector data defines objects, polygons, and other involved units so that they can be displayed or analyzed based upon their associated attributes. Raster data sets are stored as a set of uniform grid cells that represent a continuous surface.

There are many different forms that data models or bases can take. Some of the more basic and widely used ones are hierarchical, sequential, and relational. Hierarchical data models can take the form of parent-child relationships or one-to-many relationships where, for the database to be successful, direct linkages between entries are necessary. Relational data models consist of relational tables where rows are records, columns define the attribute relation, and each cell has a particular attribute value.

Management of geographic databases is very important to help ensure the accuracy of the data, the integrity of the database, and the reliability of the analysis done using a particular database. Included in this management of the database is establishing data standards, definitions, quality, and maintenance schedules. Administrators of digital geographic databases are also charged with controlling access and use of the database, backing up the database and creating and maintaining a data directory or dictionary. The data directory, also referred to as metadata, is essential to the management of a geographic database. Information included in a data directory is the definition of the particular entities, attributes and valid values. Data directories also describe the various entities and attributes when necessary, along with describing the accuracy, completeness, or other data characteristics that are important the integrity of the database.

In a world where most cellphones can determine their location, geography is important for any task that depends on where the user is physically. For example, the user may want to find a nearby restaurant — searching the web for all restaurants anywhere in the world would be absurd. Conceptually, what a location-dependent restaurant finder application might want to do is:

1. Get the current latitude and longitude from the phone hardware
2. Send a query to a restaurant database to find restaurants within X miles of the current location
3. Display the options to the user on the phone

This type of task isn't limited to the user's current location. Suppose you're flying from Boston to Paris, and want to find a hotel near the business you're going to visit. For that application, the hotel-finder application might:

1. Get the address of the business being visited
2. Convert the address to latitude and longitude
3. Send a query to a hotel database to find hotels within X miles of the specified location
4. Display the options to the user on whatever device or computer he is using

Aerodrome Mapping Database

An AMDB is a Geographic Information System (GIS) based representation describing the topography of an airport. The geometry of aerodrome features is modelled by points, lines and polygons.

Additional information characterising the features and their functions is stored as attributes. Examples of attributes are name or identifier, quality parameters and surface type. Vertical information, height and elevation, is also stored in the form of attributes for some features.

Aerodrome mapping data is provided for the following feature classes:

- Runways
- Helipads
- Taxiways
- Aprons
- Vertical structures
- Construction areas
- Water
- Hotspots
- Aerodrome surface lighting (optional)
- Quality data

These feature classes are subdivided into 33 separate feature types. For instance, the feature class "taxiways" is composed of the featuretypes taxiway element, taxiway shoulder, taxiway guidance line, taxiway intersection marking, taxiway holding position and frequency area.

AMDB Generation

In most cases electronic aerodrome data is already available, for instance in the form of engineering plans like CAD data. As AMDBs follow the defined data structure according to the RTCA/EUROCAE Standards DO-272B/ED-99B and DO291A/ED-119A, existing data can usually only be used as the basis of the data set. It needs to be converted to an AMDB, which is compliant with these standards. In doing so, it has to be ensured that the AMDB meets the requirements of RTCA/EUROCAE standards as well as of those of ICAO, especially the data quality requirements (accuracy, resolution and integrity) and the electronic data chain.

An effective method for acquisition of missing data elements or new data is photogrammetry, which is a remote sensing technology, in which geometric properties about objects are determined in form of coordinates from photographic images.

For AMDB generation aerial images or high resolution satellite images with a spatial resolution of 1 m and less are suitable. Special geometrically corrected and geo-referenced imagery allows the digitising of two-dimensional aerodrome mapping data. For AMDBs with elevation and height values, a special photogrammetry software is required for the determination of three-dimensional coordinates.

The database is a physical implementation of the Aeronautical Information Conceptual Model (AICM). In order to be compliant with the RTCA/EUROCAE Documents DO-272B/ED-99B and

DO291A/ED-119A, Avitech has extended the AICM model with additional AMDB attributes, which are not part of the original AICM model.

The AICM/AIXM 5.1 based AMDB can be used with a full AIXM 5.1 solution or in conjunction with an AIXM 4.5 solution.

The database is fully temporal, that means it is able to handle the current version of the data together with the past and future versions and display a complete history of all changes.

The Aeronautical Information Exchange Model format (AIXM 5.1) and ESRI Shapefile format are used as AMDB import and export formats. The ESRI Shapefile format is a popular GIS format used for AMDB generation, editing, processing and visualisation.

AMDBs can be used for a variety of operations. Their intended use is primarily improving the user's situational awareness and supporting pilots and air traffic controllers in surface navigation and flight operations, and as a result, increasing safety and operational efficiency. Examples of AMDB applications are position and route awareness for navigation support and traffic awareness to perform conflict and runway incursion detection and alerting.

In addition to this, further applications such as aerodrome management and vehicle operations benefit from AMDBs.

Digital Cadastral Database

Countries in the world that have maintained some form of cadastral mapping are now proceeding to a full digital representation of these cadastral maps with the ultimate aim of having country wide coverage at an accuracy level consistent with current technology in Geographic Information Systems (GIS) and surveying. GIS have shown the economic and information benefits of integrating the data sets and functions of the mapping, land titling and land management institutions. Also in each of these countries the current status and future directions of digital cadastral databases (DCDB) in any juristidiction depends very much on the historical origin of the cadastral systems and the land related institutional structures (or current restructuring). In contrast to the varied paths to the present digital cadastral maps of any country the current problems and required solutions are surprisingly similar world wide.

In contrast to the small to medium scale topographic data sets, DCDBs are very dynamic since they are tied to daily changes in the cadastral framework through the subdivision and the land titling processes. A very significant feature of the Australian digital cadastre is its full digital coverage across all the states of Australia. Each state in Australia maintains its own state DCDB and for a number of historical, political and economic reasons each has developed DCDBs that have a number of significant differences and similarities. In general the DCDB of each of the Australian states is a digital spatial database which has been derived from paper maps, it gives a graphical representation of the cadastral fabric and plays no technical or legal role in cadastral surveying or the land titling process. At present the digital cadastral database of the whole of Australia is a one time compilation of discrete and disparate data sets of each Australian states (PSMA, 1996).

The processes for updating and upgrading DCDBs are gaining considerable attention world-wide as GIS users recognise the importance of the currency, quality and content of the DCDB that underpins their GIS application. The problems that face each of the individual Australian states to satisfy both small and national GIS customers and the improvement strategies being discussed and adopted are not unlike those faced world wide by custodians of digital cadastres.

The Development of State DCDB's in Australia

In Australia the activities that relate to the capture and maintenance of a complete state digital cadastre have been the responsibility of a number of different state government bodies. The Departments of Surveyors General administered the Crown land as well as managing the surveying infrastructure and topographic mapping. State Land Surveyors Boards maintained the quality assurance function for cadastral surveys. The Land Titles Offices (LTO) for each state has had the responsibility for the titles and deeds associated with all freehold lands and have usually been responsible for examining cadastral survey plans of these lands. Land Titles Offices have specifically only ever been concerned with individual land transactions with the result that the Australian states lacked a complete cadastral map which could be used for land administration purposes (Williamson, Chan and Effenberg, 1997).

The large digital mapping data sets have their origin in the early to mid 1980's. The typical technique for establishing computerised cadastral maps or state DCDBs has been to fit the best available cadastral survey plans together onto a topographic base map using control surveys and physical features as control in a graphic 'rubber sheeting' technique. In rural areas the source maps were generally at a scale of 1:25000, while the metropolitan areas were sourced from paper map scales ranging from 1:5000 down to 1:200. In urban areas, field surveys and survey control have been used to a much greater extent with typical scales being usually 1:500- 4,000. The resulting paper cadastral map base was then digitised to establish the DCDB.

The derivation of DCDBs from paper cadastral maps means that cadastral boundaries are represented at a graphical accuracy. Large variations in the accuracy of the graphical coordinates of boundaries are possible with the accuracy of the scaled boundary coordinates in most Australian state and territory systems being about ~1 mm at map scale (eg. ~2.5 m at a map scale of 1:2,500).

Most digital cadastral maps in urban areas were initially prepared by utility authorities responsible for water, sewerage and drainage. Melbourne Water Corporation (Victoria) and Sydney Water Corporation (NSW) originally captured and managed their state's metropolitan cadastral map. These represented a large portion of the land parcels in the states of Victoria and NSW; collectively these two states have by far the majority of legal land parcels in Australia (Wan and Williamson 1995).

The mid 1980's through to early 1990's saw the capture of a digital cadastre become the key mandate for each of the Australian states (Mooney and Grant, 1997). For Victoria and NSW the cadastral data sets of the water boards for the cities of Melbourne and Sydney became the basis for each state's mandated state DCDB. The DCDBs were therefore both state and metropolitan focused, with each state generating somewhat different cadastral data sets in response to individual state priorities. As a result there are broad disparities in the quality and content of Australia's current digital cadastral map base.

Maintenance of the Cadastral Data Set

The dataflow for the incremental update of the DCDB starts with the surveyor whose deposited plans of subdivision must ultimately be correctly incorporated into the state's DCDB and eventually these changes need to be distributed to the customers of the DCDB. Figure depicts what is essentially the insertion of a digital cadastral database maintenance process into what is currently a paper based transfer of cadastral plans to the LTO by the surveyor and from the LTO to the custodian of the state DCDB's.

This post land registration model relies on the fact Land Titles Offices will capture all activity which will impact on the cadastral fabric and its attributes. Most states in Australia maintain a DCDB that is current within 5 to 10 working days of the plan being registered (OGDC, 1995 and PSMA, 1996).

Data Transfers to DCDB

The surveyor is increasingly able to offer both greater accuracy for his plans and to offer them in a digital format. Some utilities require that subdivisions of greater than say 10 parcels must be submitted in digital form It is this submitted plan in whatever format it is required to be, that is eventually used to update the each states DCDB.

To ensure the integrity and completeness of the DCDB other data sources such as government gazettes, parish plans, etc, are consulted to capture all boundary data. Some states quote up to 40% of current cadastral update activity is outside the current subdivision process (PSMA, 1996). While the cadastral activity that is not required to pass through the LTO is generally easily identified, it is often difficult to capture the related survey data.

The model clearly indicates that to include the proposed or prebuilt subdivision plan information in the DCDB would require that the information be sourced at planning stage from either the surveyor, the councils or a utility. Victoria with its close association with the metropolitan water utility, has a proposed layer for its metropolitan jurisdiction, complete with digital lodgement of subdivisions of greater than ten lots.

Data Transfers from DCDB

The custodians of the DCDB have been able to offer customers a complete copy of the DCDB in a number of digital, vendor specific and specific standard formats for some years. The DCDB is usually held in a tile or file based format and the distribution in this block format is also currently possible. The frequency of the update delivery is also flexible but in the absence of a proposed layer, customers must take account of the time required the for deposited plan to complete the LTO approval registration process.

The distribution of cadastral updates to the utilities is very much on an individual contractual arrangement. At one end of the spectrum a national utility requesting multi state coverage requires a third party to join the data sets from a number of states. In contrast for the Victorian Metropolitan DCDB, updates are able to be delivered to Melbourne Water, on a weekly basis, a bulk replacement of any of the 2,800 predefined tiles that have changed (Hesse and Jacoby, 1995).

UPDATE of state DCDB

Since the initial capture phase, the task of updating the state DCDBs to ensure that all new legal subdivisional activity is recorded and that the existing DCDB is complete, has occupied a majority of the resources of the custodians of the state DCDBs (Effenberg and Williamson, 1996).

The general method for the spatial update is to enter an image of the survey plan graphics using coordinate geometry techniques. This survey accurate graphics may or may not be stored for future upgrade purposes. Survey information is then distorted within DCDB's accuracy in this area to minimise alterations to existing cadastre. The majority of update tasks for each state, that originate from the subdivision process, are for one and two lot subdivisions.

Issues

The currency of each state DCDBs with registered plans is not presently reflected in the supply of updates to customers, while negotiable, the supply of updates to individual customers is typically six or twelve months. The custodians of the DCDB are not on the planning referral list and indeed with 177 councils in NSW and 70 odd in Victoria the information flow would be overwhelming, particularly when planning iterations are taken into account. In the absence of planning data within the DCDB (proposed layer, lodged layer, prebuilt layer) many customers such as councils and utilities, who are on the planning referral list, maintain their own portions of the DCDB and use updates from the custodian as a check on their own maintenance procedures.

In regard to planning data the concern by councils and utilities is twofold; one, as they already have the data well before the custodians of the DCDB it is easier to enter it at that time for planning purposes, and two, they are able to consider any changes to the cadastral fabric and its impact on the spatial components of their assets. This association of assets to the cadastre is a very real problem for the users of the DCDB, particularly if shifts to the cadastral boundaries occur. The result is that there are a number of large portions of each state's DCDB whose maintenance is duplicated, but state custodians however have the responsibility to maintain the state's entire cadastre.

The custodians of the DCDB offer customers a complete or 'refresh' copy of the DCDB for their area of interest. The DCDBs are also usually held in a tile or file based format and the distribution in this block format is obviously possible. Customers are increasingly requesting incremental delivery of updates for extents and frequencies of their own choosing and with internet technology, the possibility of delivering the cadastral product in more flexible ways.

Pricing is always an issue, especially in the current climate where government authorities must operate on a commercial footing. In addition to paying for the supply of updates, customers of the DCDB are also supporting a very viable industry that customizes and converts DCDB's to their specific software platform and data structure.

Improvement Strategies

In recognition of the dual role of councils and utilities as both customers and originators of the cadastral data the custodians of the DCDB are maintaining or are moving to capture a proposed plan of subdivision as part of the DCDB data model. This is a clear recognition of the fact that once the subdivision plans have been registered at the LTO, the planing authorities have already utilised or incorporated the data. It is seen that the lack of this information is the main cause for the fragmentation and duplication of the state DCDBs. If this cadastral planning information can be supplied in a timely and economical method then a more homogeneous state cadastre can be achieved.

The presence of a planning cadastral data in the cadastral model assumes that the survey planning information can be quickly accessed by the custodians of the DCDB and that this information is accessible to the DCDB customers within a time period of about 24 hours. Digital lodgement is seen as one solution to make the surveyors information more accessible and also to minimise the number of times the cadastral subdivision survey information is printed to hard copy and again entered into another computer system.

NSW and to a lesser extent Victoria have implemented the concept of mirror sites for either internal customers or specific customer research projects. Here an external database site has all of the cadastral updates for a day applied as an overnight batch process. While this is efficient for a small number of sites who maintain similar or predetermined hardware and software, it is not a viable total solution state wide. The technology of the internet offers the promise of some real data distribution and collection solutions and a number of projects are under way in Australia in this area.

As the new data models with unique feature identification to point level are implemented, incremental updates become possible but also introduce their own set of problems not the least of which is a transfer standard. Queensland has for some time offered incremental updates on the basis of an 'xy' change file in its own specific format (PSMA, 1996)

There is no doubt that the problem of spatial data associativity will remain a major problem until the accuracy of state DCDBs are upgraded to the level demanded by customers with critical underground assets. In the meantime many of these customers see a business need to update and maintain their own cadastral products.

Digital Cadastral UPGRADE

The issues relating to upgrades have their origin in the historical development of the individual digital cadastres but not surprisingly the issues of accuracy, adjustment, topology and cadastral data content are similar world wide.

Issues

The Australian Capital Territory with a comparatively small cadastre of just over 100,000 parcels, already has a cadastre accurate to ~0.03m having originally developed their DCDB directly from survey information. Other state DCDBs representing the other 10 million or so land polygons would only have some specific portions of their state DCDB approaching this accuracy.

The outlying suburban areas and to some extent the rural areas represent, for all states, the major requirement to upgrade the accuracy of the DCDB where the original capture was from 1:25,000 maps with errors of up to ~25 metres and where the rural areas are now being subdivided for fringe metropolitan housing.

GIS users in Australia are now increasingly demanding greater accuracy. Ten years ago these GIS users were happy with DCDBs having a graphical accuracy (about ~2-4m in urban areas to ~20m or more in rural areas). These same users are now demanding coordinate accuracy of the order of ~0.3m in urban areas. (Williamson and Hunter 1996)

While the surveyor may be able to supply coordinated and digital measurements the current methodology for updating the existing cadastre means that these measurements are lost when the new subdivision is entered and 'fitted' to the existing cadastral fabric.

The origin of these state DCDBs from the utilities in the last decade means that a number of issues have or need to be addressed:

- The separate development of DCDBs from the topographic data sets means that they are not aligned with topographic features.
- The CAD origin of the DCDB means a lack of topology.
- The DCDB's were property rather than parcel based.
- The cadastre contains only boundary information and no building footprints.
- The DCDB does not contain some property attributes considered essential in other cadastral data set such as land use.
- Street addressing is not fully implemented and in rural areas a street addressing program is only now being implemented.

These upgrade strategies are still only on a state by state basis and no standard data model has been adopted to more easily generate national data sets. A move to dual ownership of data sets at state boundaries is one method of beginning to formulate a national capability. The reality is that there are currently eight different cadastral systems in Australia, all with different standards.

In addition there is a real need to agree on meta data for DCDB accuracy. At present the most widely used and consistent meta data about the accuracy of portions of each state DCDB is the original map sheet from which the cadastre was sourced.

The fact that Australia has a full digital cadastral coverage of its states and territories can not be understated. This Australia wide availability of digital cadastral data has without doubt been a major contributing factor in the growth of GIS in Australia. The liability of this first generation cadastral data set is that there are significant difficulties in the update and upgrade of a graphically accurate digitised DCDB.

The future challenge for the custodians of the state DCDBs is to upgrade the DCDB to a level consistent with current technology and practises. In addition the challenge is to maintain a current and homogeneous cadastral data set for Australia. This will involve considerable utilization of current technology to gather and deliver timely and cost effective updates of the DCDB to fully meet the daily needs the data originators and customers of the state DCDBs.

Barrow Area Information Database

The Barrow area of northern Alaska is one of the most intensely researched locations in the Arctic and the Barrow Area Information Database (BAID,) tracks and facilitates a gamut of research, management, and educational activities in the area. BAID is a cyberinfrastructure (CI) that details much of the historic and extant research undertaken within in the Barrow region in a suite of interactive web-based mapping and information portals (geobrowsers). The BAID user community and target audience for BAID is diverse and includes research scientists, science logisticians, land managers, educators, students, and the general public. BAID contains information on more than 12,000 Barrow area research sites that extend back to the 1940's and more than 640 remote sensing images and geospatial datasets. In a web-based setting, users can zoom, pan, query, measure distance, save or print maps and query results, and filter or view information by space, time, and/or other tags. Additionally, data are described with metadata that meet Federal Geographic Data Committee standards. Recent advances include the addition of more than 2000 new research sites, the addition of a query builder user interface allowing rich and complex queries, and provision of differential global position system (dGPS) and high-resolution aerial imagery support to visiting scientists. Recent field surveys include over 80 miles of coastline to document rates of erosion and the collection of high-resolution sonar data for bathymetric mapping of Elson Lagoon and near shore region of the Chukchi Sea. A network of five climate stations has been deployed across the peninsula to serve as a wireless net for the research community and to deliver near real time climatic data to the user community. Local GIS personal have also been trained to better make use of scientific data for local decision making. Links to Barrow area datasets are housed at national data archives and substantial upgrades have been made to the BAID website and web mapping applications to include the public release of a new multi-temporal Imagery Viewer that allow users to interact with and compare imagery of the Barrow area from 1949 to present.

Capabilities

BAID-IMS can be used to mine information about research locations, project descriptions, photographs and contact information.

Metadata that meets the standards of the Federal Geographic Data Committee (FGDC) is available (or under development) for many data layers in BAID-IMS. Data that is considered unrestricted can be downloaded at the Arctic System Science (ARCSS) Data Coordination Center (ADCC) at the National Snow and Ice Data Center (NSIDC) located at University of Colorado at Boulder.

BAID-IMS is linked to the Circumarctic Environmental Observatories Network and efforts to develop an Arctic Spatial Data Infrastructure.

Technical Details

BAID-IMS is hosted at the University of Texas at El Paso (UTEP). The application was designed with ESRI's ArcIMS 9.2 software. BAID-IMS currently integrates geospatial data from SDE databases, shapefiles, GeoTiffs and ERDAS Imagine files. Testing is underway to migrate the application to an ArcGIS Server Web Mapping Application. Dell blade server technology is employed to host the IMS application and associated database. Bulk purchases agreements and educational discounts available to UTEP have been leveraged in the acquisition of computer hardware, software, mapping grade GPS units and digital cameras in support of this effort. Support for Open Geospatial Consortium (OGC) standards are enhanced with each upgrade to the application. An OGC Web Mapping Service (WMS) and a KML for use in Virtual Globes, such as Google Earth are also planned for release in 2008.

European Soil Database

The European Soil Database (ESDB) is an important source of data from many other data and services are derived. For instance, the "European Soil Database v2 Raster Library" contains raster (grid) data files with cell sizes of both 1km x 1km and 10km x 10km for a large number of soil related parameters. The 10km x 10km rasters are in the public domain access and allow expert users to use the data for instance to run soil, water and air related models. The 1km x 1km rasters are available after a prior registration. The grids fit with ideas from INSPIRE to develop "nested" systems for reporting and updating European (soil) data at different scales, according to a hierarchy of grids with a common point of origin and standardized location and size of grid cells.

The European Soil Database (ESDB) contains four discrete datasets which are documented in the ESDB v2.0 CD-ROM:

- the Soil Geographical Database of Eurasia at scale 1:1,000,000 (SGDBE)
- the Pedotransfer Rules Database (PTRDB)
- the Soil Profile Analytical Database of Europa (SPADBE)
- the Database of Hydraulic Properties of European Soils (HYPRES)

The Soil Geographical Database of Eurasia at scale 1:1,000,000 (SGDBE), version 4 beta

The Soil Geographical Database of Eurasia provides 1:1,000,000 coverage of soil distribution and properties for the Eurasian region. Soil classification is based on the terminology of the F.A.O. legend for the 1:5 million scale Soil Map of the World. The area covered comprises:

Albania, Austria, Belarus, Belgium, Bosnia and Herzegovina, Bulgaria, Croatia, Cyprus, Czech Republic, Denmark, Estonia, Finland, France, FYROM (Former Yugoslav Republic of Macedonia), Germany, Greece, Hungary, Iceland, Ireland, Italy, Latvia, Lithuania, Malta, Moldova, The Netherlands, Norway, Poland, Portugal, Romania, Russian Federation, Serbia, Slovakia, Slovenia, Spain, Sweden, Switzerland, Ukraine and the United Kingdom. Work is ongoing to incorporate soil data for other Mediterranean countries: Algeria, Egypt, Jordan, Lebanon, Morocco, Palestine, Syria, Tunisia and Turkey.

Potential uses include spatial data query, extraction and thematic mapping, and soil spatial data layer input to agro-environmental models (e.g. soil erosion risk, yield forecasting).

The database contains a list of Soil Typological Units (STU). As well as the soil name, each unit is described by variables (attributes) representing important properties of the soil profile - e.g. depth, texture, stoniness, water regime. At a 1:1 million scale, it is not feasible to delineate the STUs, so these are grouped into Soil Mapping Units (SMU) for mapping purposes. Each SMU represents a soil association, comprising an integrated pedological system within the landscape. In the STU table, each Soil Mapping Unit has a number of attributes. For the values of some attributes a measure of the 'confidence' in the value is given (high, moderate, low or very low). Maps are also available, giving views of the geographical database. Generally, there is one "thematic map" per STU table attribute and two accompanying maps expressing the purity and confidence level. It is important to note that the thematic maps are "dominant value" maps.

A more detailed description of the database is given in the SGDBE, while the SGDBE Dictionary explains the data structures (in the Arc/Info GIS software environment) and SGDBE Attricod provides a list of the reported attributes with their accepted values.

The database is currently managed using the ArcGIS® Geographical Information System (GIS) software system and associated relational databases.

The Pedotransfer Rules Database (PTRDB), Version 2.0

This report presents a method to translate data stored in the SGDB database to data needed for environmental purposes. The method is based on the concept of a pedotransfer function. Due to the qualitative nature of the SGDB data, these functions are simple tables and are called pedotranfer rules.

Pedotransfer rules link the standard soil characteristics to more complex properties (for practical purposes), such as hydro-dynamic properties. The rules appear in a standardized format which facilitates their use and management. These are based on expert judgement, mainly qualitative, and assume that a due weight is given to the confidence level of individual inferred attributes. The results provided by the application of these rules are therefore only qualitative estimates. At the 1:1,000,000 scale it is difficult to provide accurate information from the few data contained in the Soil Database, so methodological limitations of the approach needs to be recognised. This is done by attaching a confidence level to each output value, which can highlight those areas for which the results are not so reliable. Likewise, rules are applied to Soil Typological Units (STU) and when their results have to be displayed as maps, purity of the Soil Mapping Units (SMU) has to be taken into account. This can be computed from an indicator of the surface percentage of STUs within each SMU which is provided in the database. A set of tools has been developed within Arc/Info to manage and use the rules database.

The application of the pedotransfer rules from the PTRDB to the SGDBE results in a table that contains for each STU a set of values for each STU. These output attributes correspond to the environmental parameters considered to be relevant for the specific task: e.g., hydrology of soil types for predicting catchment response to rainfall and standard percentage of run-off; location and sensitivity of wetlands; soil buffering capacity for predicting soil susceptibility; ecosystem and surface water deposition; vulnerability of ground -and surface- water to pollution by agrochemicals and farm waste; soil erosion potential, etc.

The Soil Profile Analytical Database of Europa (SPADBE), version 2.1.0.0

The Soil Profile Analytical Database (SPADBE) is designed to enhance information about soils in the SGDBE. It contains soil profile characterisations with physical and chemical analysis. Because of problems of data harmonisation and confidentiality, two different kinds of profile data are provided:

- *Measured soil profile data* are provided where available, relating to georeferenced soil profiles that have been described and sampled in the field, and analysed in the laboratory. For measured profiles, the analytical methods are coded and missing values identified; for estimated profiles the analytical methods are defined to facilitate trans-national comparisons.
- *Estimated soil profile data* are based on expert judgement and represent the modal conditions for the STU.

The database includes standard soil property data for each horizon. Attributes include:

- parent material
- soil texture
- bulk density
- root depth
- soil water properties
- organic matter content
- structure
- pH
- cation exchange characteristics
- total nitrogen content
- calcium carbonate content
- calcium sulphate content
- electric conductivity
- cation exchange capacity and exchangeable bases

The data are provided as a set of spreadsheets (Proforma I). Since the gathered data were quite heterogeneous, hardly any attempt has been made to harmonize the data which are delivered as received. A table that holds the estimated profiles (Proforma II) is also available. An attempt has been made by the data provider or the data compiler to link each profile to an existing STU; however some profiles could not be linked. Note that estimated profiles are not geo-located.

The Database of Hydraulic Properties of European Soils (HYPRES), version 1.0

A major obstacle to the wider application of water simulation models is-

The HYPRES database was developed to overcome the lack of easily accessible and representative data on soil hydraulic properties. It contains information on a total of 5521 soil horizons. Each soil horizon was allocated to one of 11 possible soil textural/pedological classes derived from the 6 FAO texture classes (5 mineral and 1 organic) and the two pedological classes (topsoil and subsoil) recognised within the 1:1,000,000 scale Soil Geographical Database of Eurasia. Then, both categorical and continuous pedotransfer functions were developed. The categorical pedotransfer functions were used in combination with the 1:1,000,000 scale Soil Database of Europa to determine the spatial distribution of soil water availability.

The HYPRES Database contains only metadata, the actual functions and function parameters. It does not contain the source data and results, as no agreement has been reached with the participating institutions regarding their distribution.

GADM

GADM is a spatial database of the location of the world's administrative areas (or adminstrative boundaries) for use in GIS and similar software. Administrative areas in this database are countries and lower level subdivisions such as provinces, departments, bibhag, bundeslander, daerah istimewa, fivondronana, krong, landsvæðun, opština, sous-préfectures, counties, and thana. GADM describes where these administrative areas are (the "spatial features"), and for each area it provides some attributes, such as the name and variant names.

Sample View of Data

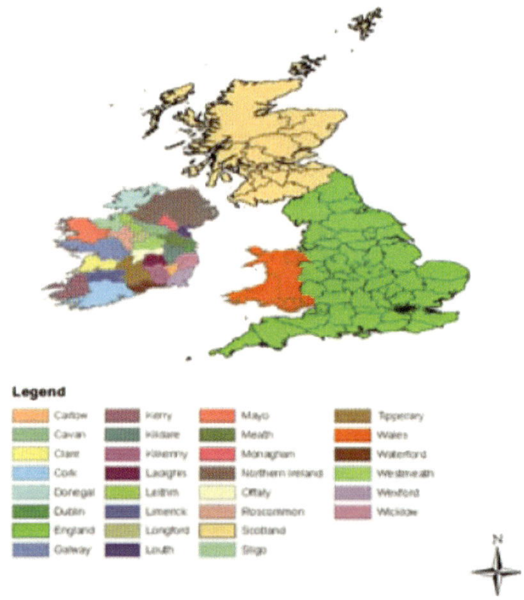

The data are available as shapefile, ESRI geodatabase, RData, and Google Earth kmz format. Shapefiles can be used for most mapping and "GIS" software. You can download a free program such as Q-GIS or DIVA-GIS. The RData files can be used in R with the 'sp' package loaded.

Data Manipulation and Storage

There are public workstations in the Geospatial Centre that provide access to ArcGIS software programs, program extensions and data manipulation utilities. Patrons may use the Gegraphic Information System (GIS) software in the library to view and manipulate the data. University of Waterloo faculty, students and staff may copy these data to other media (for example: CD, DVD or Zip disc) for use elsewhere.

Data Output

These files may be exported from ArcGIS in a choice of bitmapped formats (also known as raster graphics) including JPG, TIFF, GIF, and BMP or in a choice of vector graphics formats including Adobe Illustrator (AI), Encapsulated Postscript (EPS), and Scalable Vector Graphics (SVG). Yet another option is PDF.

6

Types of GIS File Formats

Encoding geographical information into a computer file requires a specific standard, which is called a GIS file format. The understanding of Raster graphics and Vector graphics is vital to the area of GIS file formats. This chapter discusses in detail some of the popular GIS file formats, such as GIS Raster formats and GIS Vector formats.

GIS File Formats

File formats define in which way the geographical features are stored. Raster and vector data are stored in different formats.

There are numerous formats available for both raster and vector data. It is important to consider the file format of GIS data because software programs rarely support all file types. If you want to use GIS data that was saved in a particular format not supported by your GIS program, you must either find a way to transform the data or simply use another GIS program.

Nearly all GIS programs have their own file format. These file formats were created to optimise the efficiency of the program itself, and were not designed to be used in other external programs. Nevertheless, most GIS programs support other formats by having functions to import and export datasets. These functions are usually well documented and standardised.

GIS Software and Common File Type Compatabilities

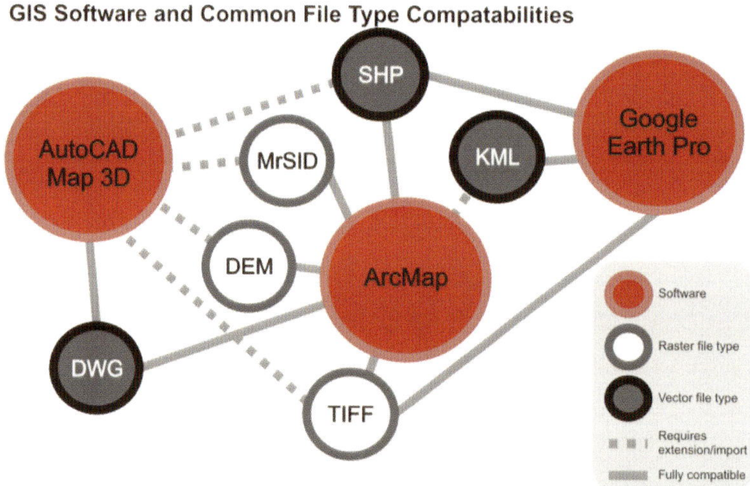

There are well over 100 GIS, CAD and Remote Sensing applications on the market. Listed here are some of the more common file formats along with associated file extension (the last 3 charcters after the period in the file name).

Raster

A raster data type is, in essence, any type of digital image represented by reducible and enlargeable grids. Anyone who is familiar with digital photography will recognize the Raster graphics pixel as the smallest individual grid unit building block of an image, usually not readily identified as an artifact shape until an image is produced on a very large scale. A combination of the pixels making up an image color formation scheme will compose details of an image, as is distinct from the commonly used points, lines, and polygon area location symbols of scalable vector graphics as the basis of the vector model of area attribute rendering. While a digital image is concerned with its output blending together its grid based details as an identifiable representation of reality, in a photograph or art image transferred into a computer, the raster data type will reflect a digitized abstraction of reality dealt with by grid populating tones or objects, quantities, cojoined or open boundaries, and map relief schemas. Aerial photos are one commonly used form of raster data, with one primary purpose in mind: to display a detailed image on a map area, or for the purposes of rendering its identifiable objects by digitization. Additional raster data sets used by a GIS will contain information regarding elevation, a digital elevation model, or reflectance of a particular wavelength of light, Landsat, or other electromagnetic spectrum indicators.

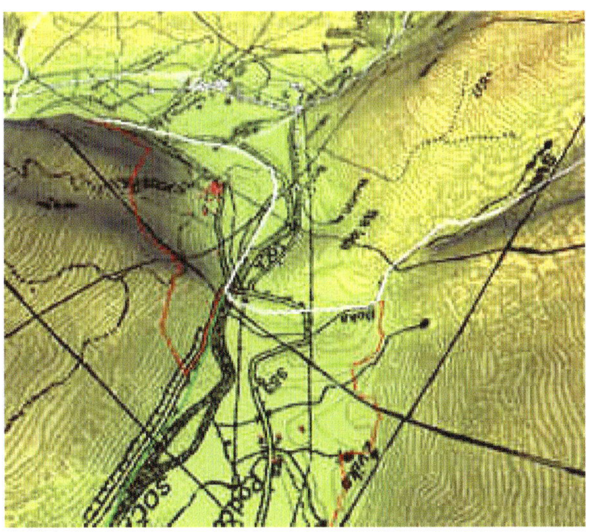

Digital elevation model, map (image), and vector data

Raster data type consists of rows and columns of cells, with each cell storing a single value. Raster data can be images raster images with each pixel (or cell) containing a color value. Additional values recorded for each cell may be a discrete value, such as land use, a continuous value, such as temperature, or a null value if no data is available. While a raster cell stores a single value, it can be extended by using raster bands to represent RGB (red, green, blue) colors, colormaps (a mapping between a thematic code and RGB value), or an extended attribute table with one row for each unique cell value. The resolution of the raster data set is its cell width in ground units.

Raster data is stored in various formats; from a standard file-based structure of TIFF, JPEG, etc. to binary large object (BLOB) data stored directly in a relational database management system (RDBMS) similar to other vector-based feature classes. Database storage, when properly indexed, typically allows for quicker retrieval of the raster data but can require storage of millions of significantly sized records.

Vector

In a GIS, geographical features are often expressed as vectors, by considering those features as geometrical shapes. Different geographical features are expressed by different types of geometry:

- Points

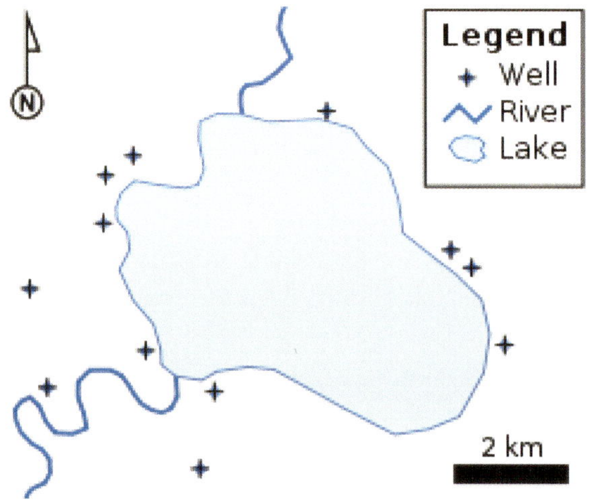

A simple vector map, using each of the vector elements: points for wells, lines for rivers, and a polygon for the lake

Zero-dimensional points are used for geographical features that can best be expressed by a single point reference—in other words, by simple location. Examples include wells, peaks, features of interest, and trailheads. Points convey the least amount of information of these file types. Points can also be used to represent areas when displayed at a small scale. For example, cities on a map of the world might be represented by points rather than polygons. No measurements are possible with point features.

- Lines or polylines

One-dimensional lines or polylines are used for linear features such as rivers, roads, railroads, trails, and topographic lines. Again, as with point features, linear features displayed at a small scale will be represented as linear features rather than as a polygon. Line features can measure distance.

- Polygons

Two-dimensional polygons are used for geographical features that cover a particular area of the earth's surface. Such features may include lakes, park boundaries, buildings, city boundaries, or land uses. Polygons convey the most amount of information of the file types. Polygon features can measure perimeter and area.

Each of these geometries are linked to a row in a database that describes their attributes. For example, a database that describes lakes may contain a lake's depth, water quality, pollution level. This information can be used to make a map to describe a particular attribute of the dataset. For example, lakes could be coloured depending on level of pollution. Different geometries can also be compared. For example, the GIS could be used to identify all wells (point geometry) that are within one kilometre of a lake (polygon geometry) that has a high level of pollution.

Vector features can be made to respect spatial integrity through the application of topology rules such as 'polygons must not overlap'. Vector data can also be used to represent continuously varying phenomena. Contour lines and triangulated irregular networks (TIN) are used to represent elevation or other continuously changing values. TINs record values at point locations, which are connected by lines to form an irregular mesh of triangles. The face of the triangles represent the terrain surface.

Advantages and Disadvantages

There are some important advantages and disadvantages to using a raster or vector data model to represent reality:

- Raster datasets record a value for all points in the area covered which may require more storage space than representing data in a vector format that can store data only where needed.
- Raster data is computationally less expensive to render than vector graphics
- There are transparency and aliasing problems when overlaying multiple stacked pieces of raster images
- Vector data allows for visually smooth and easy implementation of overlay operations, especially in terms of graphics and shape-driven information like maps, routes and custom fonts, which are more difficult with raster data.
- Vector data can be displayed as vector graphics used on traditional maps, whereas raster data will appear as an image that may have a blocky appearance for object boundaries. (depending on the resolution of the raster file)
- Vector data can be easier to register, scale, and re-project, which can simplify combining vector layers from different sources.
- Vector data is more compatible with relational database environments, where they can be part of a relational table as a normal column and processed using a multitude of operators.
- Vector file sizes are usually smaller than raster data, which can be tens, hundreds or more times larger than vector data (depending on resolution).
- Vector data is simpler to update and maintain, whereas a raster image will have to be completely reproduced. (Example: a new road is added).
- Vector data allows much more analysis capability, especially for "networks" such as roads, power, rail, telecommunications, etc. (Examples: Best route, largest port, airfields connected to two-lane highways). Raster data will not have all the characteristics of the features it displays.

Non-spatial Data

Additional non-spatial data can also be stored along with the spatial data represented by the coordinates of a vector geometry or the position of a raster cell. In vector data, the additional data contains attributes of the feature. For example, a forest inventory polygon may also have an identifier value and information about tree species. In raster data the cell value can store attribute information, but it can also be used as an identifier that can relate to records in another table.

Software is currently being developed to support spatial and non-spatial decision-making, with the solutions to spatial problems being integrated with solutions to non-spatial problems. The end result with these flexible spatial decision-making support systems (FSDSSs) is expected to be that non-experts will be able to use GIS, along with spatial criteria, and simply integrate their non-spatial criteria to view solutions to multi-criteria problems. This system is intended to assist decision-making.

Popular GIS File Formats

Raster Formats

- ADRG – National Geospatial-Intelligence Agency (NGA)'s ARC Digitized Raster Graphics.
- Binary file – An unformatted file consisting of raster data written in one of several data types, where multiple band are stored in BSQ (band sequential), BIP (band interleaved by pixel) or BIL (band interleaved by line). Georeferencing and other metadata are stored one or more sidecar files.
- Digital raster graphic (DRG) – digital scan of a paper USGS topographic map.
- ECRG – National Geospatial-Intelligence Agency (NGA)'s Enhanced Compressed ARC Raster Graphics (Better resolution than CADRG and no color loss).
- ECW – Enhanced Compressed Wavelet (from ERDAS). A compressed wavelet format, often lossy.
- Esri grid – proprietary binary and metadataless ASCII raster formats used by Esri.
- GeoTIFF – TIFF variant enriched with GIS relevant metadata.
- IMG – ERDAS IMAGINE image file format.
- JPEG2000 – Open-source raster format. A compressed format, allows both lossy and lossless compression.
- MrSID – Multi-Resolution Seamless Image Database (by Lizardtech). A compressed wavelet format, allows both lossy and lossless compression.
- netCDF-CF – netCDF file format with CF medata conventions for earth science data. Binary storage in open format with optional compression. Allows for direct web-access of subsets/aggregations of maps through OPeNDAPprotocol.
- RPF – Raster Product Format, military file format specified in MIL-STD-2411.
    - ◇ CADRG – Compressed ADRG, developed by NGA, nominal compression of 55:1 over ADRG (type of Raster Product Format).
    - ◇ CIB – Controlled Image Base, developed by NGA (type of Raster Product Format).

Vector Formats

- AutoCAD DXF – contour elevation plots in AutoCAD DXF format (by Autodesk).
- Cartesian coordinate system (XYZ) – simple point cloud.
- Digital line graph (DLG) – a USGS format for vector data.
- Esri TIN - proprietary binary format for triangulated irregular network data used by Esri.
- Geography Markup Language (GML) – XML based open standard (by OpenGIS) for GIS data exchange.
- GeoJSON – a lightweight format based on JSON, used by many open source GIS packages.
- GeoMedia – Intergraph's Microsoft Access based format for spatial vector storage.
- ISFC – Intergraph's MicroStation based CAD solution attaching vector elements to a relational Microsoft Access database.
- Keyhole Markup Language (KML) – XML based open standard (by OpenGIS) for GIS data exchange.
- MapInfo TAB format – MapInfo's vector data format using TAB, DAT, ID and MAP files.
- National Transfer Format (NTF) – National Transfer Format (mostly used by the UK Ordnance Survey).
- Spatialite – is a spatial extension to SQLite, providing vector geodatabase functionality. It is similar to PostGIS, Oracle Spatial, and SQL Server with spatial extensions.
- Shapefile – a popular vector data GIS format, developed by Esri.
- Simple Features – Open Geospatial Consortium specification for vector data.
- SOSI – a spatial data format used for all public exchange of spatial data in Norway.
- Spatial Data File – Autodesk's high-performance geodatabase format, native to MapGuide.
- TIGER – Topologically Integrated Geographic Encoding and Referencing.
- Vector Product Format (VPF) – National Geospatial-Intelligence Agency(NGA)'s format of vectored data for large geographic databases.

Grid Formats (for Elevation)

- USGS DEM – The USGS' Digital Elevation Model.
- GTOPO30 – Large complete Earth elevation model at 30 arc seconds, delivered in the USGS DEM format.
- DTED – National Geospatial-Intelligence Agency (NGA)'s Digital Terrain Elevation Data, the military standard for elevation data.

- GeoTIFF – TIFF variant enriched with GIS relevant metadata.
- SDTS – The USGS' successor to DEM.

Other Formats

- Dual Independent Map Encoding (DIME) – A historic GIS file format, developed in the 1960s.
- Geographic Data Files (GDF) — An interchange file format for geographic data.
- GeoPackage (GPKG) – An standards-based open format based on the SQLite database format for both vector and raster data.
- Well-known text (WKT) – A text markup language for representing feature geometry, developed by Open Geospatial Consortium.
- Well-known binary (WKB) – Binary version of Well-known text.
- World file – Georeferencing a raster image file (e.g. JPEG, BMP).

GIS Raster File Formats

ECW (File Format)

ECW or the enhanced compressed wavelet file extension is a type of image file format compression wherein it basically stores information captured by the used of satellites and aerial cameras or georeaders placed in space. This file type is known to be used in capturing aerial pictures via satellites for the development of geolocation and mapping databases. ECW files are popular among developers and users of Geographic Information System and thus, there are a lot of applications that can be utilized to view files with.ecw format.

Earth Resource Mapping pioneered the use of enhanced compressed wavelet and later owned by ERDAS (Earth Resource Data Analysis System). ECW is an ideal image format for large image file because of its ability to achieve large compression rates that other formats have in difficulty.

ECW files can achieve compression rates between 1:10 and 1:100, making them ideal for compressing large images.

Applications that support the ECW file extension for images include Adobe Photoshop 4.0 and 7.0, AutoCAD, Bentley MicroStation, ERDAS IMAGINE and other software from the company, ArcGIS, and others.

The ECW extension is also used for waveset format files that store articulation and sample data used by the Ensoniq software MIDI synthesizer of sound cards equipped with specific chips. Those sound cards include the Ensoniq AudioPCI and various Sound Blaster cards from Creative Labs, which now owns E-MU Systems, the original developer of Ensoniq software.

Properties

Map projection information can be embedded into the ECW file format to support geospatial applications.

Image data of up to 65,535 bands (layers or colors) can be compressed into the ECW v2 or v3 file format at a rate of over 25 MB per second on an i7 740QM (4-cores) 1.731 GHz processor using v4.2 of the ECW/JP2 SDK. Data flow compression allows for compression of large images with small RAM requirements. The file format can achieve typical compression ratios from 1:2 to 1:100.

The ECW Protocol (ECWP) is an efficient streaming protocol used to transmit ECW and JPEG2000 images over networks, such as the Internet. ECWP supports ECWPS for private and secure encrypted streaming of image data over public networks such as the Internet.

There is a very fast read-only SDK supporting ECW and JPEG2000 which is available for no charge for desktop implementation for Windows, Linux and MacOSX. A read / write SDK can be purchased for desktop and server implementations for Windows, Linux and MacOSX). A full functioning server implementation (using ECW, JPEG2000, ECWP and JPIP) is available within the PROVIDER SUITE of the Power Portfolio (formerly IWS) license. A previous version of the SDK (3.3) is available in open source, and can be used for non-Microsoft operating systems, such as Linux, macOS or Android.

USGS DEM

The USGS DEM standard is a geospatial file format developed by the United States Geological Survey for storing a raster-based digital elevation model. It is an open standard, and is used throughout the world. It has been superseded by the USGS's own SDTS format but the format remains popular due to large numbers of legacy files, self-containment, relatively simple field structure and broad, mature software support.

DEM Level

A USGS DEM can be classified into one of four levels of quality. This is due to the multiple methods of data collection, and certainty in the data.

DEM level	DEM series	Production process	Comments
Level 1	7.5-minute 30-minute	Created by autocorrelation or manual profiling from aerial photographs	May be derived from a wide variety of sources using any of a number of USGS approved techniques and instrumentation
Level 2	All series	Created from digital line graph (DLG) contours or equivalent, from any USGS map series up to 1:100000 scale	The accuracy and data spacing are intended to support computer applications that analyze hypsographicfeatures to a level of detail similar to manual interpolations of information from printed source maps
Level 3	7.5-minute	Created from DLG that have vertically integrated with all categories of hypsography, hydrography, ridge line, break line, drain files, and all vertical and horizontal control network	Requires a system of logic to interpret the data

Level 4	All series	Created from electronic (non-photogrammetric) imaging sensor systems, either active (e.g., radar or lasar) or passive (received radiant energy)	The process includes all additional post-processing required to bring the DEM grid to its final form

Format Structure

The USGS DEM format is a self-contained (single file) set of ASCII-encoded (text) 1024-byte (1024 ASCII chars) blocks that fall into three record categories called A, B, and C. There is no cross-platform ambiguity since line ending control codes are not used, and all data including numbers is represented in readable text form. There is no known binary analogue of the format, although it is common practice to compress the files with gzip.

Floating-point numbers are encoded using Fortran scientific notation, so C/C++programs need to swap the "D" exponent-indicating character with "E" when parsing (and vice versa when writing).

A record fields hold the origin, type, summary statistics and the measurement systems used by the profiles. The A record appears once as the file's header, the C record also appears once as the trailer, and multiple B records (called profiles) comprise the elevation data. A and C records each fit within one block, but a single B record typically requires multiple blocks. When such block-spanning occurs, data are shifted to start cleanly on each block boundary. A records also come in "old" and "new" flavors, because the USGS added several fields to the A record. One of the key items is the quadrangle, which is a set of four terrestrial coordinates describing the four-sided polygon enclosing the area of interest.

Starting Col	Ending Col	Description
0	134	Descriptive Name of the represented area
150	155	?
156	161	?
162	167	UTM Zone number
529	534	Unit of resolution of ground grid (0=radian;1=feet;2=metre;3=arc-second)
535	540	Unit of resolution Elevation (1=feet;2=metre)
546	569	Easting of the South West corner
570	593	Northing of the South West corner
594	617	Easting of the North West corner
618	641	Northing of the North West corner
642	665	Easting of the North East corner
666	689	Northing of the North East corner
690	713	Easting of the South East corner
714	737	Northing of the South East corner
738	761	Minimum elevation found in this file
762	786	Maximum elevation found in this file
816	827	Resolution per grid cell East – West
828	839	Resolution per grid cell North – South
858	863	Number of columns

B records (profiles) are a variable-length longitudinal column of raster elevations that start at a specified location. They are some multiple of 1024 bytes long and contain a small header summarizing the profile. The elevations are contiguous; breaks or other discontinuities are expressed using "void" elevations of value -32767. Each elevation is described as a six-character readable integer occupying a fixed location in a block. The profile header only appears in the first block, so subsequent blocks hold more elevation values. When reading the DEM file from first byte to last, one reads the profiles as columns from west to east. The elevations within a profile run from south to north. The variable-location and variable-length nature of profiles stems mainly from the use of the UTM (Universal Transverse Mercator) ground reference system. Since measurements within UTM employ fixed distances (e.g., 30 meters between elevation samples), the quadrangle must slightly distort to map such locations onto the spherical Earth. This distortion usually manifests as a rotated square, hence the elevation columns near the east and west edges start more northward and contain fewer samples.

C records contain root-mean squared error (RMSE) quality control data, using ten six-character integer fields.

MrSID

MrSID (Multi-resolution Seamless Image Database) is an image compressor, viewer, and file format for extremely large raster graphics images. MrSID, created by LizardTech, works by putting together hundreds of small image tiles into one large seamless image that can be compressed and decompressed with little or no degradation.

One problem previously associated with raster images was the long amount of time it took to open very large files. The creators of MrSID overcame that problem by giving the user the ability to decompress just the portion of the image they wanted to view. The user can move from one part of the image to another quickly without having to wait for the entire image to decompress.

Another problem associated with massive raster images was the large amount of storage space they required. MrSID uses a patented "wavelet" technology first developed at the National Research Laboratories at Los Alamos to achieve a compression ratio of 20:1 for greyscale images and 50:1 for full color images. Now, a satellite photo from space that once required the storage space of forty CD-ROMs, could be compressed with MrSID and stored on one CD-ROM.

The wavelet technology used by MrSID relies on advanced mathematical algorithms both to compress an image and build the viewer. Since the image remains geometrically accurate after compression, it can be geo-referenced before compression or overlaid with other referencing data. Because MrSID (.sid) files are a binary MIME type, they can be served over the Internet.

Common Uses

Fingerprints

According to the Open Source Geospatial Foundation (which releases GDAL), MrSID was developed "under the aegis of the U.S. government for storing fingerprints for the FBI."

Other Uses

In a 1996 entry for the R&D 100 Awards, LANL identified other uses for the format: "it can be used as an efficient method for storing and retrieving photographic archives; it can store and retrieve satellite data for consumer games and educational CD-ROMs; and it is well suited for use in vehicle navigation systems. Moreover, MrSID holds promise for being used in image compression and editing for desktop publishing and nonlinear digital video software."

For certain downloadable images (such as maps), American Memory at the Library of Congress began using MrSID in 1996; in January 2005 it also began using JPEG 2000. Depending on image content and color depth, compression of American Memory maps is typically better with MrSID, which on average achieves a compression ratio of approximately 22:1 versus the 20:1 achieved with JPEG 2000.

Software

LizardTech offers a software package called GeoExpress to read and write MrSID files. They also provide a free web browser plug-in for the Microsoft Windows operating system. (A Macintosh OS version of this viewer, introduced in 2005, was discontinued.) Most commercial GIS software packages can read some versions of MrSID files including those from GE Smallworld, ESRI, Intergraph, Bentley Systems, MapInfo, Safe Software, Autodesk, with ERDAS IMAGINE being able to both read and write MrSID files. GeoExpress can also generate JPEG 2000 (.jp2) data. When combined with LizardTech's Express Server,.sid and.jp2 data can be served quickly to a variety of GIS applications and other client applications either through direct integrations or via WMS.

There is no open source implementation of the MrSID format. Some open source GIS systems can read MrSID files, including MapWindow GIS and those based on GDAL. The Decode Software Development Kit (SDK) is made available as a free download from developer.lizardtech.com. This enables the capability to implement MrSID reading capability in any application.

Some image editing and management software systems can also read MrSID files, including XnView and IrfanView.

Technology

Encoding

MrSID technology uses lossless wavelet compression to create an initial image. Then the encoder divides the image into zoom levels, subbands, subblocks and bitplanes. After the initial encoding, the image creator can apply zero or more optimizations. While 2:1 compression ratios may be achieved losslessly, higher compression rates are lossy much like JPEG-compressed data.

Decoding

MrSID uses selective decoding meaning that the decoder does not have to decode the entire file to view a specific zoom level, image quality or scene for example.

Esri Grid

A grid is a raster data storage format native to Esri. There are two types of grids: integer and floating point. Use integer grids to represent discrete data and floating-point grids to represent continuous data.

Attributes for an integer grid are stored in a value attribute table (VAT). A VAT has one record for each unique value in the grid. The record stores the unique value (VALUE is an integer that represents a particular class or grouping of cells) and the number of cells (COUNT) in the grid represented by that value. For example, if 50 cells have a value of 1 representing a forest, then the VAT would show a VALUE = 1 and COUNT = 50 for each of the 50 cells.

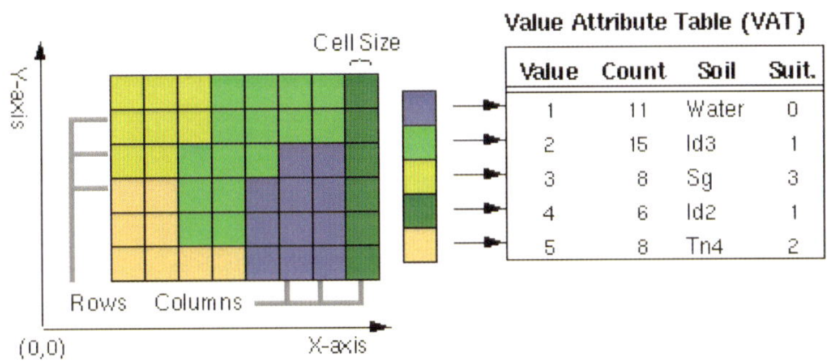

Floating-point grids do not have a VAT because the cells in the grid can assume any value within a given range of values. The cells in this type of grid do not fall neatly into discrete categories. The cell value itself is the attribute that describes the location. For example, in a grid that represents elevation data in meters above sea level, a cell with a value of 10.1662 indicates that the location is about 10 meters above sea level.

The range of data values that can be stored as grid values are as follows:

- Floating-point grids can store values from -3.4 x 1038 to 3.4 x 1038.
- Integer grids can store values from -2147483648 to 2147483647 (-231 to 231-1).

For integer grids, this information applies only to the VALUE item. An integer grid may have other INFO items added to its VAT whose range of values depends on the item definition.

The coordinate system of a grid is the same as that of other geographic data. The rows and columns are parallel to the x- and y-axes of the coordinate system. Since each cell within a grid has the same dimension as other cells, the location and area covered by any cell is easily determined by its row and column. The coordinate system of a grid is thus defined by the cell size, the number of rows and columns, and the x,y coordinate of the upper left corner. Grids also carry additional information, such as the coordinate system associated with the grid.

File Formats

ESRI ArcInfo ASCII Grid

Output format created by the GRIDASCII command (from ArcInfo Workstation) and by the Raster to ASCII tool in ArcGIS for Desktop. ESRI_ASCII_grid is an interchange format, used primarily for exchange with other programs. The file contents represents a single measure for each cell in a rectangular grid. Any coordinate system can be used to define the grid, including the widely used angular measurements with arc-seconds as units. Cells must be "square" in the chosen unit since cell size is defined by a single parameter.

The file begins with header information that defines the properties of the raster, such as the cell size, the number of rows and columns, and the coordinates of the origin of the raster. The header keywords can be in upper or lower case. The header information is followed by cell data; numeric values are ordered by column within row. Within each row, values are delimited by spaces and rows are separated by a carriage return. Either integer or floating point numbers can be used for the individual data values. The spatial location of the raster is specified by the location of the lower left cell, and by either the center of the lower left cell or the lower left corner of the lower left cell. Row 1 of the data is at the top of the raster, row 2 is below row 1, and so on.

- NCOLS *xxx*
- NROWS *xxx*
- XLLCENTER *xxx* or XLLCORNER *xxx*
- YLLCENTER *xxx* or YLLCORNER *xxx*
- CELLSIZE *xxx*

  [the length of one side of a square grid cell in same units as coordinate system used to locate the lower left center]
- NODATA_VALUE *xxx*
- Row 1 data
- Row 2 data
- etc.

Cells without actual values can be assigned a *NoData* value. The default value for *NoData* is -9999 but a different value can be specified in the header as shown in example below.

A short example:

- NCOLS 4
- NROWS 6
- XLLCORNER 378923
- YLLCORNER 4072345
- CELLSIZE 30
- NODATA_VALUE -99
- -99 -99 5 2
- -99 20 100 36

- 3 8 35 10
- 32 42 50 6
- 88 75 27 9
- 13 5 1 -99

Various discussions suggest that scientific notation may be used for the numeric values.

ESRI ArcInfo Grid

ESRI ArcInfo Grid (ESRI_grid), also known as ArcGrid, is a raster file format developed by ESRI to contain information about geographic space in a grid. A grid defines geographic space as an array of equally sized square grid points (also known as cells) arranged in rows and columns. Grids are useful for representing geographic phenomena that vary continuously over space, and for performing spatial modeling and analysis of flows, trends, and surfaces such as hydrology. Each grid cell, referenced by its x,y coordinate location, stores a numeric value that represents a geographic attribute (such as elevation or surface slope) for that unit of space. In the ESRI_grid format this numeric attribute may be a 32-bit signed integer or a 32-bit (single precision) floating point number.

The native binary ESRI Grid format comprises a number of component files. These files are stored within an ArcInfo workspace, which may incorporate several grids and other geospatial datasets, such as coverages. An ArcInfo workspace contains one INFO subdirectory and a separate subdirectory for each ESRI_grid. The INFO directory holds several files that maintain information about each grid and other datasets in the workspace. The name of an ESRI_grid cannot start with a number or use spaces, and cannot be longer than 13 characters (a multiband grid is allowed up to 9 characters). Each GRID subdirectory contains several component files that store geographic location and attribute data for the corresponding grid. Not all of them are necessary for the ESRI_grid to display. Each component file is stored in one of three storage types: INFO format ASCII format, or binary format.

In a GRID directory, the following tables and files are found:

- dblbnd.adf (aka bnd.adf) -- The boundary file contains the minimum and maximum x,y coordinates for an ESRI_grid. The boundary is a rectangle that encompasses the cells of a grid; it is stored in map coordinates. All grid BNDs are stored in double precision. This file will have a corresponding.nit and.data files in the INFO directory of an ESRI_grid. This information is stored in INFO format.
- hdr.adf -- The header file contains information about the ESRI-grid's cell resolution, type (integer or floating point), compression, and tile information. This information is in binary format.
- log -- The log file contains a history about the creation of and alterations to an ESRI_grid. This information is in ASCII format.
- prj.adf -- The projection file stores the known map projection information of the ESRI_grid. It is optional. This information is in ASCII format.

- sta.adf -- The statistics (minimum, maximum, mean, and standard deviation) of the cell values are stored in the sta.adf file. Values in the statistics are floating point values. They should be not changed using INFO commands. This file will have a corresponding.nit and. data files in the INFO directory of an ESRI_grid. More information about sta.adf files can be found in What is the file structure of an Arc/INFO Grid? This information is in INFO format.

- vat.adf -- The value attribute table (VAT) stores the the attribute data associated with the zones of the grid. The vat.adf can only exist for integer grids. This file will have a corresponding.nit and.data files in the INFO directory of an ESRI_grid. More information about vat.adf files can be found in What is the file structure of an Arc/INFO Grid? This information is in INFO format.

- w001001.adf -- Stores the value of each cell in the first (base) tile of the grid. Subsequent tiles will be numbered sequentially. This information is in binary format.

- w001001x.adf -- Stores the index for the blocks in the tile. If the grid has more than one tile, the value and index.adf file names will increment correspondingly. This information is in binary format.

- Also included in an ArcInfo Workspace is an INFO subdirectory containing files that maintain information about the ESRI_grid and other associated datasets in the workspace. In an INFO directory, the following tables and files are found:

- arcNNNN.dat (where N is any integer) -- Stores the relative path to the grid data file (e.g., "../soils/vat.adf"). This information is stored in INFO format.

- arcNNNN.nit -- Stores file structure and field definition information of the grid. For any particular grid, the NNNN values for.dat and.nit should match. This information is stored in INFO format.

- arc.dir -- Contains the data dictionary for the grid.

Two types of grids can be found in the ESRI_grid, integer and floating point. Integer grids are used to represent discrete data, i.e., that which has known and definable boundaries. An integer grid is used to determine where a data object begins and where it ends, e.g., a lake in a landscape. Floating point grids represent continuous data, i.e., data whose boundaries cannot be precisely defined; instead, the boundaries are measured in terms of a relationship to a fixed point such as elevation measured from sea level.

Grids are implemented using a tiled raster data structure in which the basic unit of data storage is a rectangular block of cells. Blocks are stored on disk in compressed form in a variable-length file structure referred to as a tile. Each block is stored as one variable-length record.

The size of the tile for a grid is based on the number of rows and columns in the grid at the time of creation. The upper limit on the size of a tile is set by the application and is very large (currently set at 4,000,000 x 4,000,000 cells). As a result, most grids used for GIS applications are automatically stored in a single tile. The spatial data for a grid is automatically split across multiple tiles if the size of the grid at the time of creation is larger than the upper limit for the size of a tile.

Grids can be combined into an ordered set of spatially overlapping grids treated as a single entity, called a grid *stack*. Grids (also known as "layers") are combined into a stack to facilitate multivariate analysis such as cluster analysis, classification, and principal component analysis. A stack has the following characteristics: a set of layers with each layer corresponding to a grid, a map extent, a cell size, a data type, and a projection stored in an ASCII file. Each layer in the grid has an index number indicating its order in the stack, all of which must be in the same workspace. The boundaries of the input layers for a grid stack can overlap exactly, partially, or not all, but the stack is considered to be only the area where the layers overlap. Computations against the stack will only be done against the overlapping area as declared by the intersection of the boundaries in the bnd.adr file. If no common area exists between the input layers, the stack will be empty and no computations will occur. As with a simple grid, the name of a grid stack cannot start with numbers, contain spaces or be longer than nine characters.

Any number of data types (real or integer) within a layer can be combined into a stack. The cell size of stack defaults to the coarsest layer in the stack. The stack is stored in the INFO directory as is information about the grids of which it is comprised. The grids of which a stack is comprised are not duplicated in a grid, but instead are referenced in the stack-related files. The directory in which the stack is located will have the following files within it:

- *.stk -- (where * indicates the user generated name of the stack) This file contains a table that stores the names of the grids that comprise the stack, their corresponding index values, and the record number of the grid. The INDEX column gives the position of a grid in the stack while the GRID column lists the grid names that comprise the stack. Other columns can be added to the stack table, but the INDEX and GRID values should not be altered unless using appropriate stack management commands found in the GRID functionality within the ArcGIS tools.
- prj.adr -- As with the description for prj.adr above, this file, when present, stores the projection information of the stack. Since a stack is treated as a single entity, all grids in a stack must be in the same projection. If the projection is unknown for all input grids in the stack, no prj.adr file is created. The projection information is used to ensure that the same geographic area is being occupied by each grid in the stack.

Grid Data Storage

A grid is stored in an ArcInfo workspace. The grid, like a coverage, is stored as a separate directory with associated tables and files that contain specific information about the grid. In an integer grid directory (originally created by ArcInfo Workstation), the following tables and files are found: the BND table, which stores the boundary of the grid; the HDR file, which stores specific information describing the grid, for example, cell resolution and blocking factor; the STA table, which contains statistics for the grid; the VAT table, which stores the attribute data associated with the zones of the grid; the log file (LOG), which monitors the activity that has occurred on the grid; and the tile file w001001.adf (q0x1y1), which stores the cell data and the accompanying index file w001001x.adf (q0x1y1x) that indexes the blocks in the tile and the LOG. (Some of these, such as the log file, may not exist if created using ArcGIS operators).

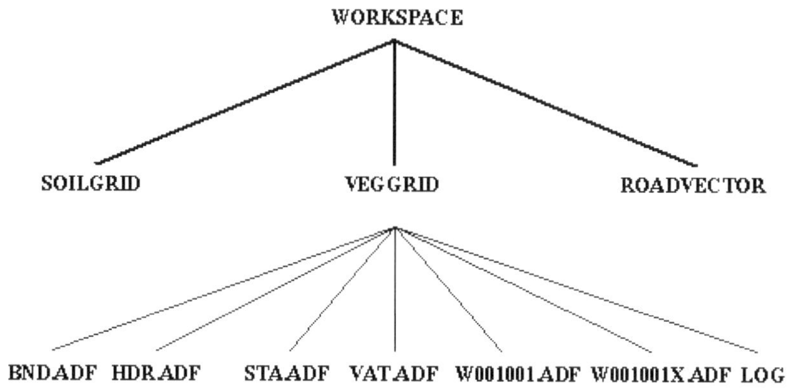

If a grid is altered, the values and information contained in the files and tables are updated immediately. The information contained in an INFO table is accessible to the user and provides information about the grid.

BND Table

A grid BND contains the boundary of the grid. The boundary is a rectangle that encompasses the cells of a grid; it is stored in map coordinates. All grid BNDs are stored in double precision.

The minimum coordinates in the BND table are for the lower left corner of the lower left cell in the grid. The maximum coordinates are for the upper right corner of the upper right cell in the grid.

HDR File

The HDR is a binary file. Information stored in the file includes the cell size, type of grid (integer or floating point), compression technique, blocking factor, and tile information.

STA Table

The STA table is an INFO table that contains statistical data about a grid. The minimum, maximum, mean, and standard deviation for the grid are stored as floating-point values in the STA table. You should not attempt to alter these values directly.

Because NoData represents an unknown value, NoData is not used in calculating the statistics in the STA table.

When a bilevel grid (containing only 0 and 1 values) is created, the STA table contains the value 0 for the mean and -1 for standard deviation. The standard deviation value -1 indicates that statistics have not been calculated for a grid.

A standard deviation value of -2 indicates that the grid contains only NoData cells.

VAT Table

The VAT is an INFO table that stores attributes associated with the zones of a grid. Only integer grids have a VAT associated with them. Every VAT has at least two items, VALUE and COUNT. The VALUE item contains integer values that are used to distinguish the characteristics of one location from the other locations in a grid. All cells that are assigned the same value contain the same characteristics and, therefore, belong to the same zone. COUNT is the number of cells in a zone.

New items can be added to the VAT. The VALUE and COUNT items should not be changed, and the VAT must be kept sorted on the VALUE item. Never add new items before VALUE or COUNT.

Cells containing NoData are not represented in the VAT.

Below is an example of a VAT:

Record	VALUE	COUNT
1	0	628872
2	1	265043
3	2	151150
4	3	3185652
5	4	79983
6	5	4782
7	6	74334
8	7	8877
9	8	1817
10	9	491
11	10	858
12	11	8770
13	12	28789
14	13	72539
15	14	3686
16	15	3932
17	16	13227
18	17	1890
19	18	1305
20	19	427286
21	20	6695

Tile Files

The w001001.adf (q0x1y1) and w001001x.adf (q0x1y1x) files store the data and the index for the first, or base tile, in a grid. The upper limit on the size of a tile is very large, and most grids are stored using a single tile. If additional tiles are used, they are automatically numbered based on their spatial relationship to the first tile. Tiles are implemented as variable-length binary files. With versions prior to ARC/INFO 7.x, these files were named q0x1y1 and q0x1y1x and still work with the current software.

LOG File

The LOG file is an ASCII file that contains information about the creation of and alterations to a grid. The LOG monitors the actions performed on the grid, but it does not contain every action performed with the grid. Since all Grid functions result in a new grid, only Grid commands, such as RENAME and COPY, can alter an existing grid and be entered into the LOG file. The LOG file can be accessed, like all ASCII files, through system commands or any text editor.

Storage Limitations

- The name of a grid is limited as follows:
- It cannot be stored using spaces.
- It cannot start with a number.
- It cannot be longer than 13 characters (a multiband grid is allowed up to 9 characters).

There is a limit to the number of files that can be stored in an INFO directory for both coverages and grids. This total is approximately 10,000. Therefore, this limits the number of grids you can store in a workspace. For example, the following lists the theoretical maximum number of grid datasets that can be stored in a single workspace directory:

- Fewer than 5,000 floating point grids, or
- Fewer than 3,333 integer grids, with VATs (fewer than 5,000 if no VATs), or
- Fewer than 10,000 grid stacks

The preceding numbers are the theoretical maximums. If you have a process that will create interim grids (and therefore files in the INFO directory) these numbers will be less. Additionally, if you are storing a mix of files, such as grids and coverages, you will store fewer.

These numbers relate to the number of files in the grid folder that store information in the INFO directory. The limit is 10,000 (9,999), but it's not the total number of files in an INFO directory, it's the number of files pointing to the files in the INFO directory. For each grid, there are two files in the grid's folder pointing to files in the INFO folder: the BND (boundary) files, and STA table (statistics) files (9999/2≈5000). When a grid has a VAT, this also points to files in the INFO directory, so the number that can be stored is reduced again (9999/3≈3333). A grid stack only has a single file which points to the INFO directory (9999/1≈9999).

Stack

A stack consists of an ordered set of spatially overlapping grids (layers) treated as a single entity for multivariate analysis. Cluster analysis, classification, and principal component analysis all work on the layers in a stack.

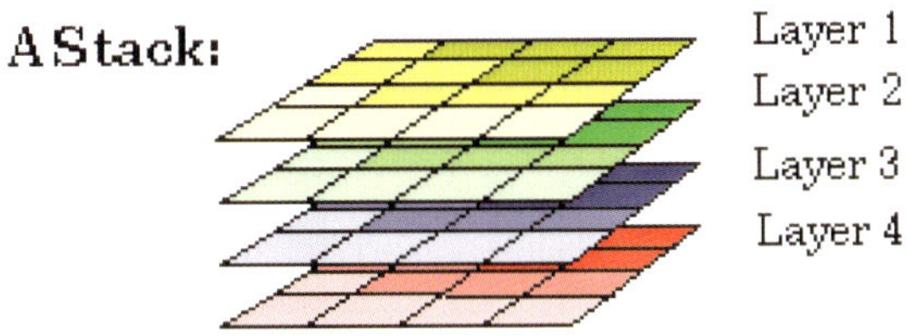

A stack has the following characteristics:

- A set of layers with each layer corresponding to a grid
- A map extent, or BND
- A cell size
- A data type
- A projection

Each layer specified in a stack has an index number indicating its order in the stack. The grids that make up a stack must be in the same workspace.

The boundaries of the input layers can overlap exactly, partially, or not at all, but only the area where layers overlap comprises the stack. The stack's BND is where the boundaries of its layers intersect. The computations of a multivariate analysis function occur on the overlapping area. If there is no common area between the input layers, the stack is empty and no computations occur.

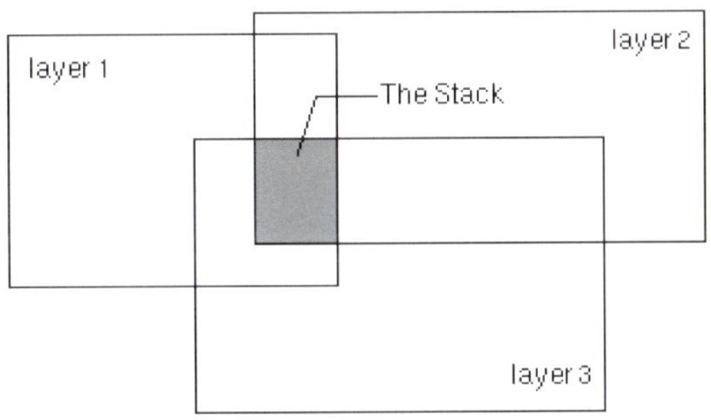

The cell size of a stack defaults to the coarsest layer in the stack.

You can combine any number of data types (real or integer) of the input grids in a stack. However, before applying a multivariate technique, you should be aware of what the values represent, whether categorical or continuous data, and the range or relative range of the values. In certain analysis, the input data type of the stack determines the data type of the output.

Projection information associated with the input grids is stored with the stack. Since a stack is treated as a single entity, all grids in a stack must be in the same projection. The projection information is used to ensure that each grid of the stack occupies the same geographic area.

Storing a Grid Stack

A stack is stored in a directory structure similar to a grid or coverage. There are two files in the stack directory: an external INFO STK table and an ASCII PRJ file. The actual grids that comprise the stack are not stored in the stack. They are ordinary grids in your workspace. That means any grid can be used in more than one stack. The STK table stores the names of the grids that comprise the stack and their corresponding index values:

GRID: LIST JER135.STK

Record	INDEX	GRID
1	1	jer1
2	2	jer3
3	3	jer5

The INDEX item gives the position of a grid in the stack, while the GRID item lists the grid names that comprise the stack. The spatial data of the input grids is not duplicated in the stack. As a result, the stack always reflects the latest version of the input grids. The STK file is as accessible as any other INFO file. You can add items for descriptive purposes, such as an item for storing the date that the data was collected, but don't use INFO to alter the values in the INDEX item or names in the GRID item. All manipulations to these items should only be performed using a variety of the stack management commands available in Grid.

The PRJ file, when present, stores the projection information of the stack:

Projection STATEPLANE

Zone	4701
Datum	NAD27
Zunits	NO
Units	FEET
Spheroid	CLARKE1866
Xshift	0.0000000000
Yshift	0.0000000000
Parameters	

If the projection is unknown for all input grids in the stack, no PRJ file is created.

The name of a grid stack cannot be stored using spaces, cannot start with a number, and cannot be longer than 9 characters.

NoData in a Grid

Every cell in a grid has a value assigned to it; however, cells without actual values can be assigned NoData on the grid representing that theme. NoData and 0 (zero) are not the same; 0 is a valid value. For this reason, NoData cells cannot be used in calculating the statistics in a grid's STA table.

GIS Vector File Formats

CityGML

CityGML is a common information model for representing 3D urban objects. It defines classes and relations for the most relevant topographic objects in cities and regional models with respect to their geometric, topological, semantic and appearance properties. “City” is broadly defined to include not just built structures, but also elevation, vegetation, water bodies, “sidewalk furniture” and more. Included are generalization hierarchies between thematic classes, aggregations, relationships between objects and spatial properties.

These thematic information types go beyond graphic exchange formats and allow users to employ virtual 3D city models for sophisticated analysis tasks in different application domains such as simulation, urban data mining, facilities management, decision support and thematic inquiries.

CityGML is an open data model and XML-based format for storing and exchanging virtual 3D city models. It is implemented as an application schema of GML3, the extensible international standard for spatial data exchange developed within the Open Geospatial Consortium (OGC) and ISO TC211.

GML3, used with other OGC standards - mainly the OpenGIS Web Feature Service (WFS) Specification - provides a framework for exchange of simple and complex 3D models. However, WFS and GML3 only establish syntactic interoperability. A GML3 document needs to be structured by the definition of an application schema that is tailored to a specific application domain. In this case, the application domain is 3D city modeling and the GML3 application schema (or profile) is CityGML.

CityGML takes advantage of other open standards and its development has proceeded in careful cooperation with other groups. For example, graphic rendering of data encoded in CityGML can be accomplished using standardized computer graphics data formats like VRML, GeoVRML, X3D or Universal 3D (U3D). To ensure European acceptance, the developers of CityGML have coordinated with EuroSDR, a spatial data research organization consisting of delegates from geographic information production organizations and research centers from 18 member states in Europe, together with participants from industry and the commercial sector.

Within the OGC, discussion of CityGML takes place within the OGC CAD/GIS/3D Interoperability Working Group. CityGML will be submitted to the OGC Technical Committee, meeting in Edinburgh, Scotland during the week of June 26-29, as a candidate OpenGIS specification for review and possible adoption by the membership. The intention is for it to become an open standard and therefore to be used free of charge.

Development of CityGML began in 2002 by the members of the Special Interest Group 3D (SIG 3D) of the Geodata Infrastructure North-Rhine Westphalia (GDI NRW) initiative in Germany. The GDI NRW has been active in the OGC since 1999. The SIG 3D is an open group consisting of more than 70 companies, municipalities, and research institutions working on developing and commercially exploiting interoperable 3D models and visualization. Another product of SIG 3D's activities is a proposed Web 3D Service (W3DS) standard, a 3D portrayal service that currently has the status of a discussion paper in the OGC (OGC Doc. No. 05-002).

Some of the entities working on and supporting CityGML are the municipalities of Berlin, Hamburg, Cologne, Düsseldorf, Recklinghausen and Leverkusen; the Ordnance Survey Great Britain; the State Mapping Agency of North-Rhine Westphalia; companies including T-Mobile, Rheinmetall Defence Electronics, Snowflake, CPA Geo-Information, GIStec and 3D Geo; and research institutions including the universities of Bonn, Potsdam, Dortmund, Applied Sciences Stuttgart, the Helmholtz Research Centre Karlsruhe, and the Fraunhofer Institute for Graphics Research in Darmstadt.

CityGML Features

CityGML represents the graphic appearance of city models and also the semantic or thematic properties, taxonomies and aggregations of digital terrain models, sites (including buildings, bridges and tunnels), vegetation, water bodies, transportation facilities, etc. The underlying model differentiates five consecutive levels of detail (LoD), where objects become more detailed with increasing LoD, both in geometry and thematic differentiation. CityGML files can - but don't have to - contain multiple representations for each object in different LoD simultaneously.

At this point, the UML diagrams, XML schema definitions, example datasets and free viewer applications are available from the CityGML homepage. Basic concepts and most of the modeling decisions are explained in a number of conference papers and presentation slides which can also be found on the website. The OGC will publish the CityGML specification document when it becomes an approved Discussion Paper.

An open source GML3 3D viewer application called Aristoteles has been developed by the Institute for Cartography and Geoinformation at the University of Bonn. It is based on Java 1.5 and Java3D. Aristoteles is not restricted to CityGML files; it can visualize arbitrary GML3 files (like the German cadastre standard ALKIS/NAS or the OS Mastermap from UK).

A free CityGML viewer application called LandXPlorer was developed by 3DGeo and the Hasso-Plattner-Institute (HPI) of the University of Potsdam. The LandXplorer CityGML Tool allows users to interactively view, import and export CityGML-based model data. For non-profit use and with limitations regarding the model size and the available functions, the LandXplorer CityGML Tool can be used for free.

CityGML in the OGC OWS-4 Testbed

The OGC's main interoperability initiative this year is the fourth OGC Web Services testbed (OWS-4). OWS-4 kicked off in June and will conclude with an interoperability demonstration in the fall of 2006. OWS-4 sponsors agreed that this initiative would advance geospatial standards in the area of CAD/GIS/BIM (building information model) integration, as well as in other areas such as sensor webs, geoprocessing workflow and geospatial digital rights management.

OWS-4 is based on a fictitious scenario that will be played out in the final demonstration. The CAD/GIS/BIM use case in the scenario involves siting and planning a temporary field hospital in an emergency situation. The commander determines that a desirable potential site is an abandoned building at the local airport. Facility planners analyze the location, including consideration of access by road and helicopter. Security is a high concern.

CityGML will be rigorously exercised and perhaps extended by the OWS-4 participants. A CAD Model Server (CMS) will be created to provide access to various kinds of CAD data, such as as-built data for buildings originally in CAD files without coordinate system information, road data in civil design applications, underground utilities CAD drawings and manholes in photogrammetric, planimetric data in a CAD file.

A Web Feature Service (implementing the OGC WFS standard) will be developed to provide client programs with access to GIS site data (terrain, orthophotos, land use), accessible as GML, and to Aeronautical Information Exchange Model (AIXM) data available for a helicopter pad.

The scenario outlines a basic course of action that will put the various servers and clients through a series of operations designed to test and demonstrate the utility of CityGML.

A virtual reality constructed from diverse kinds of data representing the built environment can benefit millions of people working in a long list of industries. As bandwidth, processing and storage keep getting faster and cheaper, lowering technical and financial hurdles, it makes more and more sense for stakeholder groups to imagine how such a shared virtual reality might benefit them.

CityGML is the result of a well-organized effort to bring together related standards through careful coordination with various standards groups and user groups. After years of development, it is now in "beta" and it will be tested in OWS-4. CityGML deserves careful review by a wide range of industry groups studying geospatial data sharing, CAD/GIS interoperability, industry foundation classes (IFCs), building information models (BIMs), and workflow and business models in building-related professions.

GeoJSON

GeoJSON is an open standard format designed for representing simple geographical features, along with their non-spatial attributes. It is based on JSON, the JavaScript Object Notation.

The features include points (therefore addresses and locations), line strings (therefore streets, highways and boundaries), polygons (countries, provinces, tracts of land), and multi-part collections of these types. GeoJSON features need not represent entities of the physical world only; mobile routing and navigation apps, for example, might describe their service coverage using GeoJSON.

The GeoJSON format differs from other GIS standards in that it was written and is maintained not by a formal standards organization, but by an Internet working group of developers.

A notable offspring of GeoJSON is TopoJSON, an extension of GeoJSON that encodes geospatial topology and that typically provides smaller file sizes.

Example

```
{
  "type": "FeatureCollection",
  "features": [
   {
     "type": "Feature",
     "geometry": {
       "type": "Point",
       "coordinates": [102.0, 0.5]
   },
   "properties": {
     "prop0": "value0"
   }
  },
  {
   "type": "Feature",
   "geometry": {
     "type": "LineString",
     "coordinates": [
```

```
        [102.0, 0.0], [103.0, 1.0], [104.0, 0.0], [105.0, 1.0]
    ]
   },
   "properties": {
    "prop0": "value0",
    "prop1": 0.0
    }
   },
   {
    "type": "Feature",
    "geometry": {
      "type": "Polygon",
      "coordinates": [
      [
        [100.0, 0.0], [101.0, 0.0], [101.0, 1.0],
        [100.0, 1.0], [100.0, 0.0]
      ]
    ]
   },
   "properties": {
    "prop0": "value0",
    "prop1": { "this": "that" }
   }
  }
 ]
}
```


Supported Software

GeoJSON is supported by numerous mapping and GIS software packages, including OpenLayers, Leaflet, MapServer, Geoforge software, GeoServer, GeoDjango, GDAL, Safe Software FME, and

CartoDB. It is also possible to use GeoJSON with PostGIS and Mapnik, both of which handle the format via the GDAL OGR conversion library. Bing Maps, Yahoo!, HERE, and Google also support GeoJSON in their API services.

The Google Maps JavaScript API v3 directly supports the integration of GeoJSON data layers as of March 19, 2014. For the Julia language a GeoJSON.jl package is available.

GitHub also supports GeoJSON rendering and Potrace GeoJSON export.

Geojson.io supports GeoJSON rendering and editing in the web browser.

TopoJSON

TopoJSON is an extension of GeoJSON that encodes topology. Rather than representing geometries discretely, geometries in TopoJSON files are stitched together from shared line segments called *arcs*. *Arcs* are sequences of points, while line strings and polygons are defined as sequences of arcs. Each arc is defined only once, but can be referenced several times by different shapes, thus reducing redundancy and decreasing the file size. In addition, TopoJSON facilitates applications that use topology, such as topology-preserving shape simplification, automatic map coloring, and cartograms.

A reference implementation of the TopoJSON specification is available as a command-line tool to encode TopoJSON from GeoJSON (or ESRI Shapefiles) and a client side JavaScript library to decode TopoJSON back to GeoJSON again. TopoJSON is also supported by the popular OGR tool as of version 1.11 and PostGIS as of version 2.1.0.

TopoJSON Schema

Given a GIS shape near coordinates latitude 0° and longitude 0°, a simple but valid and complete topojson file containing all metadata, *Polygon*, *LineString*, *Point* elements, *arcs* and *properties* is defined as follows:

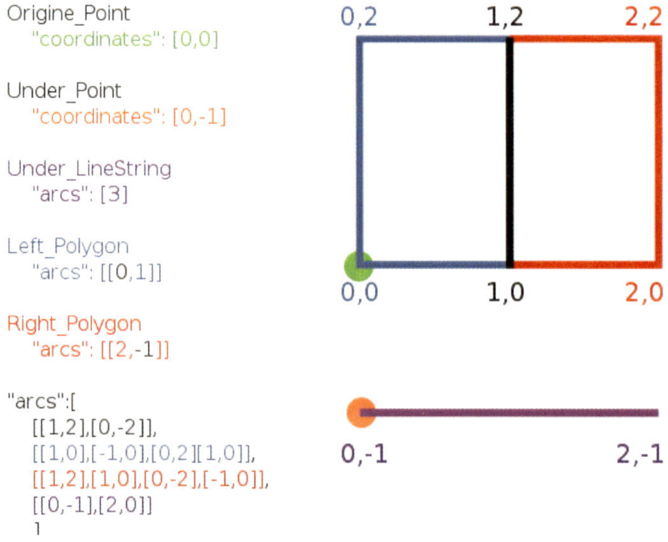

Topojson Shapes

```
{
  "type":"Topology",
  "transform":{
    "scale": [1,1],
    "translate": [0,0]
  },
  "objects":{
    "two-squares":{
      "type": "GeometryCollection",
      "geometries":[
        {"type": "Polygon", "arcs":[[0,1]],"properties": {"name": "Left_
        Polygon" }},
        {"type": "Polygon", "arcs":[[2,-1]],"properties": {"name":
        "Right_Polygon" }}
    ]
  },
  "one-line": {
    "type":"GeometryCollection",
    "geometries":[
     {"type": "LineString", "arcs": ,"properties":{"name":"Under_Line-
     String"}}
    ]
  },
  "two-places":{
    "type":"GeometryCollection",
    "geometries":[
     {"type":"Point","coordinates":[0,0],"properties":{"name":"Origi-
     ne_Point"}},
     {"type":"Point","coordinates":[0,-1],"properties":{"name":"Under_
     Point"}}
```

```
      ]
    }
  },
    "arcs": [
      [[1,2],[0,-2]],
      [[1,0],[-1,0],[0,2],[1,0]],
      [[1,2],[1,0],[0,-2],[-1,0]],
      [[0,-1],[2,0]]
    ]
 }
```

MapInfo TAB Format

Sometimes called the MapInfo Native format or the MapInfo TAB format, the MapInfo dataset format is a proprietary format developed for use with MapInfo software products. A MapInfo dataset consists of a number of physical files. Two files are required for textual and numeric data without graphic objects. The table structure in ASCII format (*.TAB) is the main file for a MapInfo table. The.TAB ASCII file is the link between all other files and holds information about the type of data file. The attribute data (*.DAT) is stored in binary format. These two files support a "browser" view of the tabular data, a view comparable to a spreadsheet.

File Components

The basic file components for a MapInfo Professional data set relate to the two basic environments for working in MapInfo; "Browser View" and "Mapper View".

As with most other GIS packages, several files are required to allow the user to open a data set for viewing within MapInfo Professional. The most basic view would be the browser view only. This environment provides storage of attribute or object data and is represented like a spreadsheet. In this simplified scenario, no geographic information is available.

Minimum files required for the basic MapInfo Professional browser environment:

- .TAB (The ASCII file that is the link between all other files and holds information about the type of data set file).
- .DAT (The file that stores the attribute data. This is a dBase III DBF file).

As an alternative to the *.DAT file, MapInfo Professional can use other data formats such as, *.TXT, *.XLS *.WK*, *.MDB (and for each Microsoft Access format the software also makes another small file). MapInfo Professional still creates a.TAB file that contains information about the data set file, and the user interacts with the TAB file only.

There may also be a third file:

- .IND (Optional index file for tabular data. This is present if any columns are indexed).

To view geographic information (the graphic representation of data) in MapInfo Professional, two additional files are required and added to the basic requirements for simply viewing data.

Minimum files required to view a map with the data previously discussed:

- .MAP (Stores the graphic and geographic information needed to display each vector feature on a map)
- .ID (Stores information linking graphic data to the database information. This contains a 4-byte integer index into the MAP file for each feature).

Each time a *.MAP file is saved by MapInfo Professional its objects are sorted into a storage sequence based on an R-tree spatial index. This optimises the read process of streaming data from disk to screen, at the cost of a relatively slow write process.

So the basic file set for viewing data and its graphic representation in vector form within MapInfo Professional requires a minimum of four files, normally the *.TAB, *.MAP, *.DAT and *.ID. If there is only textual information with no graphic objects, then a minimum of two files is needed: *.TAB and *.DAT.

While a data set is being edited by a user, temporary files are created by MapInfo Professional. The presence of any of these files locks the data set and prevents it from being edited by any other users. When the changes are saved, the temporary files are deleted by the software. In certain situations such as a power failure while editing is in progress, the temporary files may not be deleted and the changes will be lost. In this case the files should be manually deleted.

Normal temporary files have the following file name extensions:

- .TDA Temporary database file
- .TIN Temporary index file
- .TMA Temporary Map File

When using a database table such as Oracle Locator or Spatial, if the data is downloaded to the local machine, the temporary file name extensions are:

- .LDA Local Temporary database file
- .LIN Local Temporary index file
- .LMA Local Temporary Map File

MapInfo Professional can also use raster images and grid surfaces as layers in a map. These are stored in their native file formats, such as *.JPG, *.TIF or *.PNG, and the software again creates a *.TAB file for each one which includes georeferencing information.

A seamless table enables a single layer in a map to be derived from multiple base data sets, such as a mosaic of raster image tiles. When MapInfo Professional creates a seamless table it makes a simple vector layer which contains a rectangle for each base data set with attribute data defining the file path to the files containing the base data. When the seamless table is opened, metadata in its *.TAB file causes the map to display not the rectangles but the contents of the base tables in their place.

SOSI

SOSI is a much used geospatial vector data format for predominantly used for exchange of geographical information in Norway.

SOSI is short for Samordnet Opplegg for Stedfestet Informasjon (literally "Coordinated Approach for Spatial Information", but more commonly expanded in English to Systematic Organization of Spatial Information).

The standard includes standardized definitions for geometry and topology, data quality, coordinate systems, attributes and metadata.

The open standard was developed by the Norwegian Mapping and Cadastre Authority. It was first published in 1987 (version 1.0). It is continuously being revised and further developed. The long term development points towards international standards (ISO 19100). This work is being done by ISO/TC211, currently chaired by Olaf Østensen with the Norwegian Mapping and Cadastre Authority.

Shapefile

The ESRI Shapefile (known here as the ESRI *Shapefile format*), stores nontopological geometry and attribute information for the spatial features in a data set. A shapefile consists minimally of a main file, an index file, and a dBASE table.

In the main file, the geometry for a feature is stored as a shape comprising a set of vector coordinates. This main file is a direct access, variable-record-length file in which each record describes a shape with a list of its vertices. In the index file, each record contains the offset of the corresponding main file record from the beginning of the main file. Attributes are held in a dBASE format file. The dBASE table contains feature attributes with one record per feature. Attribute records in the dBASE file must be in the same order as records in the main file. Each attribute record has a one-to-one relationship with the associated shape record.

The shapefile format can support point, line, and area features. Area features are represented as closed loop, double-digitized polygons.

Instances of the Shapefile format have often been used as a data exchange format from ESRI formats to non-ESRI applications. The format is most useful for writing simple features and attributes quickly as there are limitations inherent in the Shapefile format related to both geometry and attributes. As outlined elsewhere in this description, these limitations may cause loss of data when using shapefiles to contain or exchange complex geometry or attributes. The Shapefile format may be used as an intermediary between data creation applications and more functionally capable GIS formats and applications, albeit with the limitations noted in the Dataset/Normal Dataset section.

The cluster of files is typically stored in the same file directory or project workspace, with all component files having the same filename (prefix) and identified by individual file extension (suffixes). Three components are mandatory: a main file that contains the feature geometry (.shp), an index file that stores the index of the feature geometry (.shx), and a dBASE table (.dbf) that stores the attribute information of features. A comprehensive list of component files follows:

- *shp* -- Main file (mandatory); a direct access, variable-record-length file in which each record describes a shape with a list of its vertices.
- *shx* -- Index file (mandatory). In the index file, each record contains the offset of the corresponding main file record from the beginning of the main file. The index file (.shx) contains a 100-byte header followed by 8-byte, fixed-length records.
- *dbf* -- dBASE Table file (mandatory); a constrained form of DBF that contains feature attributes with one record per feature. The one-to-one relationship between geometry and attributes is based on record number. Attribute records in the dBASE file must be in the same order as records in the main file.
- *sbn* -- Part 1 of spatial index for read-write instances of the Shapefile format. If present, essential for correct processing.
- *sbx* -- Part 2 of spatial index for read-write instances of the Shapefile format. If present, essential for correct processing.
- *atx* -- Created by ArcView 3.x for each instance of the Shapefile format or dBASE attribute index created in ArcCatalog. ArcView GIS 3.x attribute indexes for shapefiles and dBASE files are not used by later versions of ArcGIS as a new attribute indexing model has been developed for shapefiles and dBASE files.
- *fbn* -- One of the files that store the spatial index of the features for instances of the Shapefile format that are read-only.
- *fbx* -- The other file (besides.fbn) that stores the spatial index of the features for instances of the Shapefile format that are read-only.
- *ain* -- One of the files that stores the attribute index of the active fields in a table or a theme's attribute table.
- *aih* -- The other file (besides.ain) that stores the attribute index of the active fields in a table or a theme's attribute table
- *ixs* -- Geocoding index for read/write shapefiles. If present, essential for correct processing.
- *mxs* --Geocoding index for read-write shapefiles (ODB format).
- *prj* -- Projections Definition file; stores coordinate system information.
- *xml* -- contains metadata, as used by ArcGIS.
- *cpg* -- An optional file that can be used to specify the codepage for identifying the character set to be used.

Shapefile Shape Format (.shp)

The main file (.shp) contains the geometry data. The binary file consists of a single fixed-length header followed by one or more variable-length records. Each of the variable-length records includes a record-header component and a record-contents component. This format should not be confused with the AutoCAD shape font source format, which shares the.shp extension.

The 2D axis ordering of coordinate data assumes a Cartesian coordinate system, using the order (X Y) or (Easting Northing). This axis order is consistent for Geographic coordinate systems, where the order is similarly (longitude latitude). Geometries may also support 3- or 4-dimensional Z and M coordinates, for elevation and measure, respectively. A Z-dimension stores the elevation of each coordinate in 3D space, which can be used for analysis or for visualisation of geometries using 3D computer graphics. The user-defined M dimension can be used for one of many functions, such as storing linear referencing measures or relative time of a feature in 4D space.

The main file header is fixed at 100 bytes in length and contains 17 fields; nine 4-byte (32-bit signed integer or int32) integer fields followed by eight 8-byte (double) signed floating point fields:

Bytes	Type	Endianness	Usage
0–3	int32	big	File code (always hex value 0x0000270a)
4–23	int32	big	Unused; five uint32
24–27	int32	big	File length (in 16-bit words, including the header)
28–31	int32	little	Version
32–35	int32	little	Shape type
36–67	double	little	Minimum bounding rectangle (MBR) of all shapes contained within the dataset; four doubles in the following order: min X, min Y, max X, max Y
68–83	double	little	Range of Z; two doubles in the following order: min Z, max Z
84–99	double	little	Range of M; two doubles in the following order: min M, max M

The file then contains any number of variable-length records. Each record is prefixed with a record header of 8 bytes:

Bytes	Type	Endianness	Usage
0–3	int32	big	Record number (1-based)
4–7	int32	big	Record length (in 16-bit words)

Following the record header is the actual record:

Bytes	Type	Endianness	Usage
0–3	int32	little	Shape type
4–	–	–	Shape content

The variable-length record contents depend on the shape type, which must be either the shape type given in the file header or Null. The following are the possible shape types:

Value	Shape type	Fields
0	Null shape	None
1	Point	X, Y
3	Polyline	MBR, Number of parts, Number of points, Parts, Points
5	Polygon	MBR, Number of parts, Number of points, Parts, Points
8	MultiPoint	MBR, Number of points, Points
11	PointZ	X, Y, Z *Optional*: M
13	PolylineZ	*Mandatory*: MBR, Number of parts, Number of points, Parts, Points, Z range, Z array *Optional*: M range, M array
15	PolygonZ	*Mandatory*: MBR, Number of parts, Number of points, Parts, Points, Z range, Z array *Optional*: M range, M array
18	MultiPointZ	*Mandatory*: MBR, Number of points, Points, Z range, Z array *Optional*: M range, M array
21	PointM	X, Y, M
23	PolylineM	*Mandatory*: MBR, Number of parts, Number of points, Parts, Points *Optional*: M range, M array
25	PolygonM	*Mandatory*: MBR, Number of parts, Number of points, Parts, Points *Optional*: M range, M array
28	MultiPointM	*Mandatory*: MBR, Number of points, Points *Optional Fields*: M range, M array
31	MultiPatch	*Mandatory*: MBR, Number of parts, Number of points, Parts, Part types, Points, Z range, Z array *Optional*: M range, M array

Shapefile Shape Index Format (.shx)

The index contains the same 100-byte header as the.shp file, followed by any number of 8-byte fixed-length records which consist of the following two fields:

Bytes	Type	Endianness	Usage
0–3	int32	big	Record offset (in 16-bit words)
4–7	int32	big	Record length (in 16-bit words)

Using this index, it is possible to seek backwards in the shapefile by, first, seeking backwards in the shape index (which is possible because it uses fixed-length records), then reading the record offset, and using that offset to seek to the correct position in the.shp file. It is also possible to seek forwards an arbitrary number of records using the same method.

Shapefile Attribute Format (.dbf)

This file stores the attributes for each shape; it uses the dBase format. An alternative format that can also be used is the xBase format, which has an open specification, and is used in open source shapefile libraries, such as the Shapefile C library.

The names and values of attributes are not standardized, and will be different depending on the source of the shapefile.

Shapefile Spatial Index Format (.sbn)

This is a binary spatial index file, which is used only by Esri software. The format is not documented by Esri. However it has been reverse-engineered and documented by the open source community. It is not currently implemented by other vendors. The.sbn file is not strictly necessary, since the.shp file contains all of the information necessary to successfully parse the spatial data.

Best Practices

Below are best practices for using shapefiles in Map Viewer.

- Choose to generalize features for web display if you have a large amount of features intended to be displayed at small scales and you have not already generalized your vertices in ArcGIS Desktop.
- Create a.zip file that contains the.shp,.shx,.dbf, and.prj files that comprise the shapefile.
- Store your shapefile directly in the root (the central directory) of the.zip archive and not in directories within the archive. If your.zip file viewer shows path information, the path should be blank.
- The name of the new layer is the same as the name of the.zip file you added. You can rename the file before adding to the map, or rename the layer once it is part of the map.
- Add standard compressed archive.zip files.
- Shapefiles that have more than 4,000 point features (or 2,000 line or polygon features) or are larger than 10 MB cannot be added directly to Map Viewer. If the file is too large, generalizing the features will often reduce the overall size and allow the shapefile to be added to the map. If that doesn't work and you have publishing privileges in your organization, you can publish your shapefile as a hosted feature layer.
- The shapefile should contain valid geometries. If you have ArcMap, you can use the Repair Geometry geoprocessing tool to correct invalid geometries in shapefiles.
- Include in the shapefile the accompanying.prj file in which the coordinate system of the data is defined. If you have ArcMap, you can define the coordinate system of a shapefile, which creates an appropriate.prj file for the shapefile.

Limitations

The following limitations apply to using shapefiles in Map Viewer:

- Compression formats other than a.zip archive are not supported.
- The shapefile must be less than 10 MB in size and have less than 4,000 point features or

2,000 line or polygon features. If you have publishing privileges in your organization, it is recommended that you publish shapefiles that exceed these limits as a hosted feature layer.

- The following features are not supported: multipatch or multipoint geometries, geometries that cross the dateline, or self-intersections in polygons. Shapefiles with these features cannot be added to a map.
- Shapefiles cannot be dragged directly onto Map Viewer; you must use the Add button in Map Viewer.

Permissions

All chapters in this book are published with permission under the Creative Commons Attribution Share Alike License or equivalent. Every chapter published in this book has been scrutinized by our experts. Their significance has been extensively debated. The topics covered herein carry significant information for a comprehensive understanding. They may even be implemented as practical applications or may be referred to as a beginning point for further studies.

We would like to thank the editorial team for lending their expertise to make the book truly unique. They have played a crucial role in the development of this book. Without their invaluable contributions this book wouldn't have been possible. They have made vital efforts to compile up to date information on the varied aspects of this subject to make this book a valuable addition to the collection of many professionals and students.

This book was conceptualized with the vision of imparting up-to-date and integrated information in this field. To ensure the same, a matchless editorial board was set up. Every individual on the board went through rigorous rounds of assessment to prove their worth. After which they invested a large part of their time researching and compiling the most relevant data for our readers.

The editorial board has been involved in producing this book since its inception. They have spent rigorous hours researching and exploring the diverse topics which have resulted in the successful publishing of this book. They have passed on their knowledge of decades through this book. To expedite this challenging task, the publisher supported the team at every step. A small team of assistant editors was also appointed to further simplify the editing procedure and attain best results for the readers.

Apart from the editorial board, the designing team has also invested a significant amount of their time in understanding the subject and creating the most relevant covers. They scrutinized every image to scout for the most suitable representation of the subject and create an appropriate cover for the book.

The publishing team has been an ardent support to the editorial, designing and production team. Their endless efforts to recruit the best for this project, has resulted in the accomplishment of this book. They are a veteran in the field of academics and their pool of knowledge is as vast as their experience in printing. Their expertise and guidance has proved useful at every step. Their uncompromising quality standards have made this book an exceptional effort. Their encouragement from time to time has been an inspiration for everyone.

The publisher and the editorial board hope that this book will prove to be a valuable piece of knowledge for students, practitioners and scholars across the globe.

Index